Spar Platforms

Technology and Analysis Methods

SPONSORED BY
Ocean and Offshore Engineering Committee
Coasts, Oceans, Ports, and Rivers Institute
of the American Society of Civil Engineers

EDITED BY
Moo-Hyun Kim, Ph.D.

ASCE

COPRI
COASTS, OCEANS, PORTS & RIVERS INSTITUTE

Published by the American Society of Civil Engineers

Library of Congress Cataloging-in-Publication Data

Spar platforms: technology and analysis methods / edited by Moo-Hyun Kim.
p. cm.
Includes bibliographical references and index.
ISBN 978-0-7844-1209-1 (pbk.) — ISBN 978-0-7844-7677-2 (e-book)
1. Offshore oil well drilling. 2. Drilling platforms. 3. Underwater drilling. I. Kim, Mu-hyon. II. American Society of Civil Engineers.
TN871.3.S684 2012
622'.33819—dc23

2012014866

Published by American Society of Civil Engineers
1801 Alexander Bell Drive
Reston, Virginia 20191
www.asce.org/pubs

ISBN 978-0-7844-1209-1 (paper)
ISBN 978-0-7844-7677-2 (e-book)
Manufactured in the United States of America.

18 17 16 15 14 13 12 11 1 2 3 4 5

Contents

Preface

The chapters of this ASCE monograph on spar technology and analysis have been collected by me as the editor and published as an initiative of the ASCE Ocean and Offshore Engineering Committee (OOEC). The committee published a similar monograph, "Tension-Leg Platform," in 1989. Many renowned experts from academia and the offshore industry contributed to this work, and I would like to thank them all. This spar monograph includes numerous up-to-date topics of wide interest to the scientific and industrial communities that will be helpful in educating practicing engineers and graduate students alike. Most of the design, experimental, and analytical methods described in this monograph are also indirectly applicable to other floating compliant platforms.

The spar platform was selected as a theme structure by the Offshore Technology Research Center (OTRC) industry board members in 1992 and followed by several years of intensive research and experimentation. As a product of this endeavor, the first spar, Oryx Neptune, was successfully installed in the Gulf of Mexico in 1996. Since then, spars have been very popular as a proven and reliable solution for oil production in deep and remote areas of the Gulf of Mexico. To date three spar concepts—(1) classic spar (a deep-draft hollow vertical cylinder), (2) truss spar (a combination of relatively shallow-draft hollow cylinder and truss structure with heave plates), and (3) cell spar (a combination of small-diameter tubes extended to a soft tank)—have been proposed, designed, and installed. As of March 2006, three classic and 11 truss spars are producing oil and gas, and two more truss spars and a cell spar are under construction. Although spar technology and relevant analysis methods are now considered to be mature, there still exist several technical challenges to be resolved, such as vortex induced motion (VIM) and its suppression, and survivability in category 4 and 5 hurricanes.

Chapter 1 summarizes the history of spar development. Chapter 2 describes the detailed design aspects of spar mooring and anchoring system. Chapters 3 through 6 explain various methodologies to analyze spar global motions, with Chapters 3 and 4 using the diffraction method and Chapters 5 and 6, the Morrison formula. Although the Morrison formula is simpler than the diffraction method, it can only be used for long-wave conditions, such as 100-year storm, and it is not suitable for operational conditions.

Chapter 7 describes various design considerations needed for spar VIM, and Chapter 8 introduces active thruster control as a possible remedy. Chapter 9 explains various aspects and considerations of spar model testing in a wave basin. Finally, Chapter 10 shows some full-scale field data obtained from an existing spar platform and compares them with the results of numerical analysis tools.

For the global motion analysis of floating systems like the spar, nonlinear time-domain simulation methods are recommended, including the coupling effects of the hull, mooring lines, and risers. The deepwater-spar experiments can be done with an equivalent truncated mooring system. However, to better represent the inertia and damping effects from the full length of actual mooring lines, the experimental numerical hybrid model testing method is recommended. In this approach the numerical model is calibrated for the truncated system and then extrapolated to simulate the full-length case. The full-scale field data, including the failure incidents during hurricane Ivan, Katrina, and Rita, are becoming available, and they seem to correlate reasonably well with numerical prediction by reliable computer programs, although more comparative studies and forensic analyses are necessary in the coming years to reach any meaningful conclusions.

The offshore industry has many useful guidelines and practice manuals for spar. This monograph supplements rather than supersedes those existing materials. Many of its chapters address selective areas associated with spars from a scientific perspective and attempt to clarify a number of unclearly explained problems. I did my best to distribute topics among chapters to cover both practical and scientific aspects of spar technology and analysis. These features should make this book valuable to designers, researchers, and graduate students of offshore engineering.

The content of this book has been prepared in accordance with recognized engineering principles and is believed to be accurate. However, the views and materials it contains are based on the personal opinions and experiences of the chapter authors. Neither the ASCE nor the affiliates of each contributor warrant the accuracy of this document.

"Joseph" M. H. Kim
Chairman (2002–2004)
Ocean and Offshore Engineering Committee

Chapter 1: Spar Floating Drilling, Production, and Storage System: History and Evolution

By John Halkyard, Sc.D., P.E., John Halkyard & Associates, Houston, Texas

Abstract

The idea of using a Spar for a drilling, production, and storage system originated in the early 1980s. After numerous studies and joint industry projects, Oryx Energy and CNG installed the first actual unit in 1996. As of this writing, 17 Spars have been installed. All of these units are in the deepwater Gulf of Mexico except for one in the South China Sea in East Malaysia. There are now three generations of Spar—the Classic, Truss, and Cell Spars.

This paper reviews the evolution of past and present Spar designs, focusing on the progression of work that ultimately lead to the application of this new concept to the oil industry. This includes the evolution from classic, deep draft caisson designs to the Truss Spar and the Cell Spar.

Introduction

This paper updates a previous review of Spar evolution presented at the 10th Offshore Symposium of the Texas Section of the Society of Naval Architects and Marine Engineers (SNAME) in Houston in February 2001. At that time three Classic Spars were in place and two Truss Spars were under contract. In the ensuing years, there has been an explosion of Spar developments, resulting in a total of 17 Spars having now been installed. (see Appendix A for a complete list). During the extreme hurricane seasons of 2004 and 2005, the Spars in deep water proved to be robust and stable platforms, even in environments that exceeded the 100-year design condition. This is a remarkable success story. The Spar's inventor, Ed Horton, has moved on to a new venture and a new concept, but his Spar legacy will remain for many years. A list of milestones in the Spar evolution is included as Appendix B.

In the Beginning

"Archimedes was right." That was Ed Horton's comment about the successful upending of the Oryx Neptune Spar in September 1996. It was the installation of the world's first production Spar platform.

The Spar was not a new concept for use offshore. The Shell/Exxon Brent Spar was installed in the North Sea in the 1970s and was used as a successful storage and offloading terminal for more than 19 years.

Fig. 1-1. Brent Spar under construction (Photograph courtesy of Quistnix)

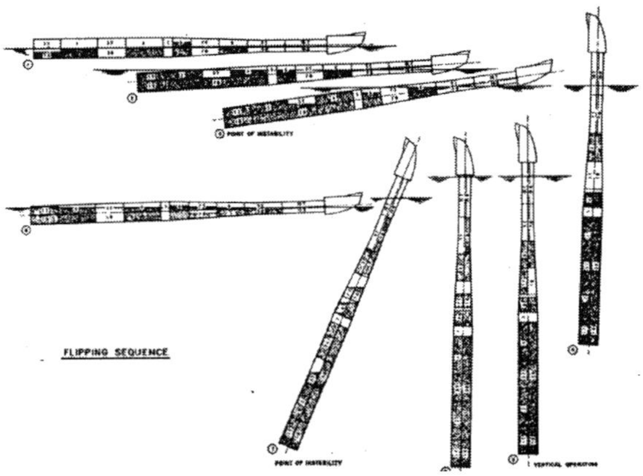

Fig. 1-2. FLIP was built in 1962 (Source: Fisher and Spiess 1963. Reproduced with permission from Journal of the Acoustical Society of America.)

The stability of the Spar design was utilized by oceanographers with the 1962 christening of the Scripps FLIP vessel.

The possibilities of using the Spar as a production platform were studied in the 1970s by some of the major oil companies, although its use as a production platform never materialized. These studies concentrated on using the Spar with wet trees.

Ed Horton was also not new to inventing new production concepts. He invented the tension leg platform (TLP) in the 1960s and established Deep Oil Technology, Inc. (DOT) in the late 1960s to develop that idea. That incarnation of DOT was subsequently purchased by Fluor Corporation and eventually dissolved. Horton realized from this experience that it was important to remain independent while developing a new idea, at least until the concept was advanced to the stage of proven feasibility and all the patent rights had been secured if possible.

Deep Oil Technology, Inc.

In 1984 Horton had recently left Global Marine Development Company and was looking for new challenges. He formed a new company, resurrecting the name Deep Oil Technology, Inc., (DOT) to pursue his latest invention—the Spar production and storage concept. Horton was familiar with FLIP and its good motion properties. He also was familiar with the Brent Spar and thought that this concept would be the ideal form for a production platform. Having invented the tension leg platform concept in the 1960s, he understood the challenges of supporting dry trees. Because Spar heave motions could be very low, he concluded that it should be possible to combine dry trees and oil storage on a Spar.

In September 1984 Horton presented the Spar concept to Arco Oil and Gas Company, which subsequently awarded DOT a study contract to analyze motions, mooring loads, and riser behavior. Horton turned to his friend and colleague, Randy Paulling, a professor of naval architecture at the University of California at Berkeley, to help estimate the motions of a Spar. Paulling had developed a suite of computer programs, which had already been successfully calibrated against TLP and semi-submersible motions. Paulling used a frequency domain program, SPLASH, to estimate Spar motions. This program showed that the heave response of a Spar decreased rapidly with the draft and reached a point of diminishing returns at a draft of about 650 ft.

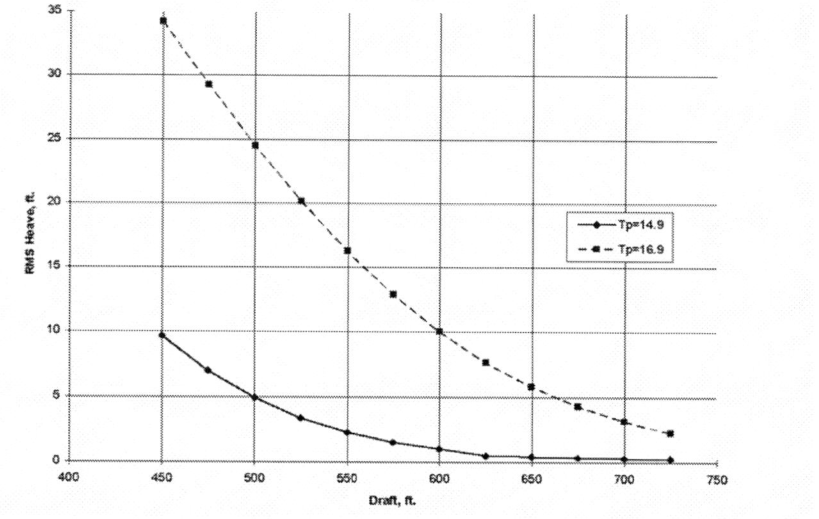

Fig. 1-3. Heave predictions with linear theory suggested 650 ft. was an optimum draft

Some means of supporting the risers while accommodating this amount of heave motion was required for the Spar to work with dry trees. Riser tensioners, such as those used on TLPs, did not have adequate stroke capacity. Horton's idea was to build a centerwell in the Spar and support the risers on steel buoyancy cans. These cans are guided in the centerwell, but the Spar is free to move vertically relative to the risers. Stroke was not a problem as long as the buoyancy cans could be constrained within the guides. That idea distinguished this production Spar from those previously considered for production.

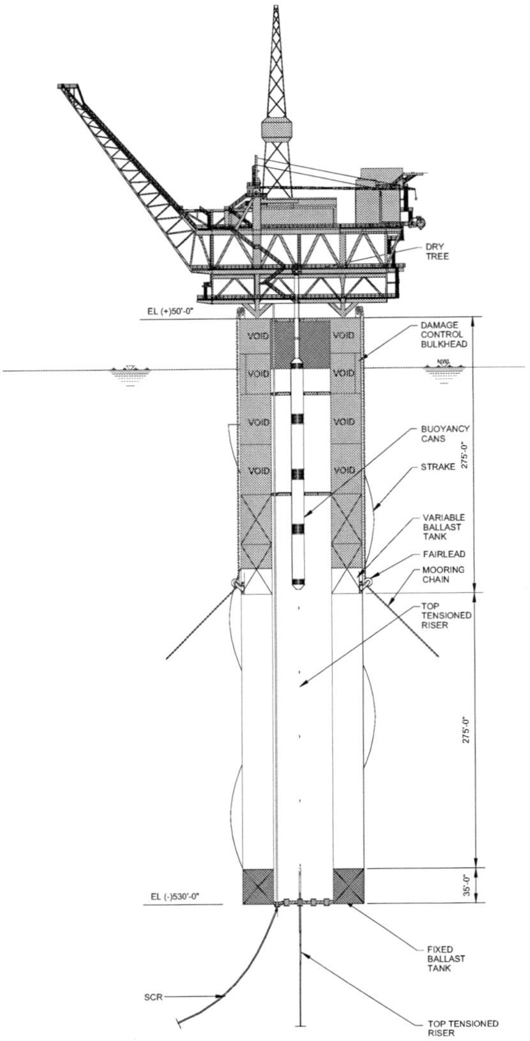

Fig. 1-4. Risers supported on buoyancy cans represented the key innovation of the Spar (Courtesy of Technip USA, Inc.)

The Arco study indicated that the Spar was a very promising concept and that further work was warranted. In September 1985 Arco commissioned an independent study with Earl and Wright in San Francisco to evaluate the feasibility of DOT's conceptual design and to develop a cost estimate. The study for Arco confirmed that DOT's design was feasible and that the cost was lower than other competing concepts. Unfortunately, shortly thereafter, Arco stopped all funding for deep water projects including the Spar.

To keep development flowing, Horton (now the sole employee of DOT) sought money from other oil companies to do some tests and design work. Questions about the Spar's behavior in the late 1980s centered on heave motions. DOT received a joint industry study contract in 1987 to investigate Spar motions. Seven oil companies—Shell, Texaco, Amoco, Arco, Sun, Elf, and Chevron—participated.

Horton again turned to Paulling to help. The University of California had a small wave and towing tank in Richmond, Calif. With some funding from Amoco, Horton and Paulling built some small Spar models and tested them in the UC Berkeley tank.

Fig. 1-5. Ed Horton with one of the first Spar models (photo by author)

Horton doubled as lab technician and wound up in the tank in his bathing suit hooking up mooring lines and instrument cables. A major concern in those days was the effect of second-order heave motions. To investigate this, Paulling devised a clever experiment consisting of a fixed vertical pipe filled with water that was open at the bottom. The water in the pipe was dynamically equivalent to a Spar; however, because its surge and pitch motions were constrained, its motions were purely heave. Additionally, Reynold's scale effects were less pronounced for the water motion inside the pipe. Excitation was due solely to the pressure fluctuations at the bottom of the centerwell. Waves were generated in the tank, and the motions of the water in the pipe were measured.

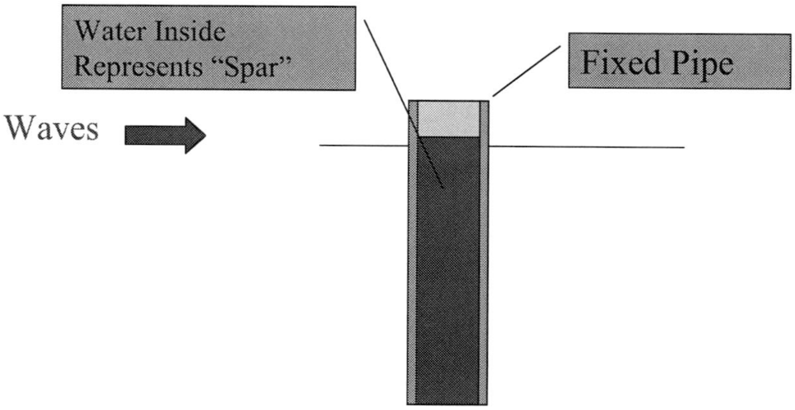

Fig. 1-6. Schematic of Paulling's early experiment to investigate Spar heave motion

Because the second-order heave forces arise from wave group effects in which the difference in the frequency of waves equals the natural heave period of the Spar, a number of experiments were conducted with two sine waves of different frequencies matching this criterion. Tests were also conducted with irregular waves corresponding to storm waves. Amoco provided a 1:280 scale model of a Spar with a centerwell, which was moored in the basin to simulate an actual Spar installation.

The tests concluded that Spar heave motion could be adequately represented by linear theory for the heave forces if quadratic damping was considered. The second-order wave excitation was found to be present but of insignificant consequence.

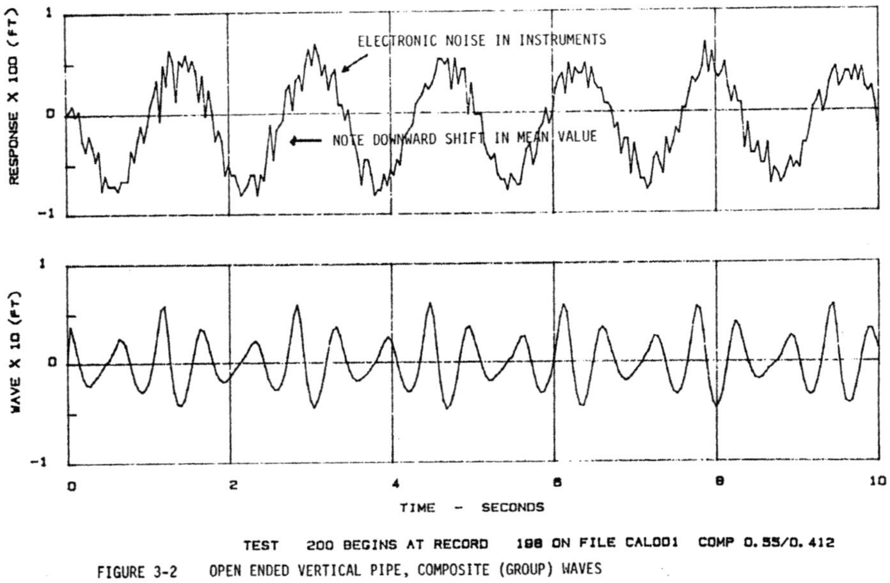

Fig. 1-7. Second-order heave response (upper plot) is shown to be < 1/100 of the wave group amplitude at resonance (Courtesy of Technip USA, Inc.)

Tension Buoyant Tower

Although the idea of the Spar was originally based on oil storage, most of the interest in the late 1980s was for a version without oil storage. Horton's answer was the tensioned buoyant tower (TBT). This Spar had tendons, similar to a TLP, risers that were tied directly to the deck, and no lateral mooring lines. The lower section of original TBT hull was a conical or stepped diameter cylinder. Later versions replaced this with a truss section.

DOT formed a joint industry study in 1988 to assess the TBT. A major part of the study was a model test conducted at the Arctec Offshore wave and towing tank in Escondido, Calif. This facility was much larger than the Berkeley tank and allowed testing at a larger scale and with more instruments.

Paulling was again called on to provide numerical predictions, and the model tests were designed to validate them. At the time I was working at Arctec Offshore and was assigned the job of managing the tests.

This time the coupled response of the risers, tendons, and TBT platform were of interest. Paulling had developed a new program in the early 1980s for the Ocean Thermal Energy Conversion (OTEC) project. This program, called COPIPE, solved the coupled equations of motion for the platform and pipe.

While the model tests successfully confirmed the analysis of the TBT in waves, it also had profound implications for future Spar designs.

One objective for the tests was to consider wave-current interaction effects. Because the Arctec basin didn't have current generation capability, an experiment was designed with the TBT tendons attached to a rigid frame, which in turn was attached to the towing carriage. Towing the assembly simulated a current acting on the TBT. In addition, waves could be generated to compound this simulated current.

The first towing tests were conducted on a Friday afternoon with all the JIP participants present. The result of the first test surprised everyone. When the TBT was towed without waves, it unexpectedly moved from side to side. Horton said later that he felt he had been bitten by a snake. There was great consternation and uncertainty over what to do next. I had seen vortex induced vibrations (VIV) of cylinders before. Pierre Beynet from Amoco (now BP) was familiar with VIV, and we all concluded that this was indeed what was happening. The consensus was to try to suppress the VIV with an arrangement of external pipes that would mimic helical strakes.

These decisions were being made late Friday afternoon. None of the participants wanted to leave the basin until this matter had been resolved and the TBT could be towed without VIV. I saw this as clearly outside the scope of our fixed-price contract, and in the interest of preserving Arctec's profits, I objected to an extended Friday night effort with all of our engineers and technicians on overtime. This did not ingratiate me with Horton or the participants who had come a great distance to see a successful test. After being convinced of the importance to the future of the TBT and indeed the Spar, I agreed to press on.

That night we tried several approaches to reducing the VIV. We tied pipes on the outside and wrapped them in a spiral pattern around the upper hull. VIV was reduced, but the hull still moved considerably. It was later speculated that the smaller diameter lower hull (as opposed to a truss section used today on the Truss Spar) might have been responsible for this motion. Finally around midnight, the participants decided to go back to the drawing board on VIV, and I was allowed to close shop.

This experience notwithstanding, the TBT seemed like a good concept for small well counts and smaller topsides, but the Classic Spar still seemed more flexible. In 1989 Chevron was looking for a solution for its Green Canyon Block 205 (GC 205) discovery. It was looking at a compliant tower for 2,500 ft. water depth. Some in Chevron's research department recognized the potential of the Spar concept and asked DOT to study it for the GC205 field. By this time, the Arctec model basin was in bankruptcy, and I was laid off. Horton offered me a job helping with the Chevron study as long as I could work for food and some equity in DOT. This was the worst job offer I have ever had, but it turned out to be a most rewarding job.

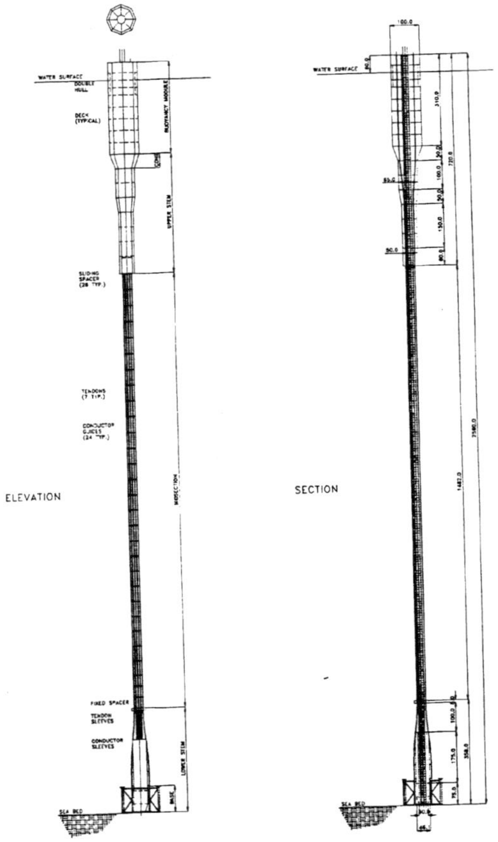

Fig. 1-8. The tension buoyant tower looked promising but never got off the ground (Source: Halkyard et al. 1991. © 1991, Society of Petroleum Engineers, Inc. Reproduced with permission of SPE. Further reproduction prohibited without permission.)

The 1989 GC 205 study was the first time the Spar concept was evaluated comprehensively for an actual Gulf of Mexico field. The study's scope included all aspects of a Spar project including configuration, motions, mooring, construction, and installation. The proposed hull form was an octagon with flat sides for simple shipyard construction. Several shipyards were approached with the idea of building all components in the United States if possible. Costs were based on construction at Avondale's yard in New Orleans, La. Analysis verified the feasibility of using a taut mooring with the Spar, which helped reduce mooring costs significantly.

While we believed the octagonal form was the proper, cost-effective shape for the hull, the TBT tests had raised a red flag regarding VIV behavior (the TBT was also an octagonal hull form). With support from Chevron, we returned to the Berkeley towing tank in March of 1990 to test an octagonal and a circular Spar shape with and without helical strakes. The Spar was moored to the tow carriage for these tests, and we simulated a current with the tow carriage again.

Irv Brooks from Chevron was responsible for managing our work on these tests. He recognized the advantages of the Spar and became a champion of the concept as a drilling and production platform for GC 205.

About the time the contract for these tests was initiated, Horton was closing a deal that would keep DOT solvent as it continued to develop and market the Spar. He had previous dealings through Global Marine with the Finnish company, Rauma-Repola (Now Technip Finland), which built more semi-submersible drilling rigs at its yard in Pori than any other yard in the world. Rolf Ohman, president of the company's U.S. office, recognized the Spar's potential. He thought Pori would be an ideal yard for Spar construction. Together with Horton they convinced the home office in Finland to enter into a marketing agreement with DOT. In exchange for exclusive world-wide marketing rights, Rauma-Repola helped finance DOT's R&D and sales efforts. It also provided an engineer, Timo Lehtinen, to work with DOT in its Irvine, Calif. offices. One of Lehtinen's first assignments was to help run the Chevron tests at Berkeley.

About this time Ed also added more people to his staff. He brought on board one of his earliest friend's from his days in Chevron, Roger Glanville. Roger is a chemical engineer, but during his career he transitioned to project management and finance. He was to become the DOT Program Manager for the first spar project. But his early responsibilities were to keep the money going out equal to or less than the money coming in. Another former associate of Ed's from Global Marine days, Dick Davies, was hired to head up the design efforts. Dick along with myself became the technical department. I brought in a former colleague of mine from my early Ocean Mining days, Atle Steen, who was an ace at doing the critical calculations of the mooring and responses. This was the group that would ultimately design the first spar to be built. We all had to work for a minimum salary in those days, but Ed made up the difference in equity in the company. This proved to be a great move for us.

These tests convinced us that the octagonal Spar would exhibit VIV behavior that could not be suppressed by usual methods. The motions and the in-line drag forces were 2.5 times higher for the octagonal Spar, with or without suppression, as they were for the circular Spar with suppression.

Research Picks Up

As the Berkeley tests began, DOT was finishing the first phase of another joint industry project it initiated to develop a theory and numerical solution for the problem of Spar VIV. This project measured the oscillatory lift and drag forces on vertical cylinders towed from a dynamometer attached to the tow carriage at the Arctec Offshore towing tank. The tests produced a large empirical database that could be used in a numerical algorithm to predict the amplitude of vortex shedding. The data included both circular and octagonal shapes, with and without helical strakes. Thus we could, in theory, predict the effectiveness of different strake configurations. Solving the non-linear VIV equations was somewhat complicated, so I enlisted the help of a colleague from college and my earlier career in ocean mining, Atle Steen. Steen also lived in San Diego, Calif., thus we were able to collaborate on testing as well as theoretical work. Following this effort, Steen joined DOT and became an integral part of the Spar development.

At the completion of these programs, Chevron's research department asked DOT to prepare a preliminary design for a GC205 Spar to be compared with a TLP also being studied. The preliminary design was based on a circular Spar shape and incorporated the knowledge of VIV and Spar motions gained from the recent tests. Paulling performed the analysis using his time domain simulation program, TDSIM6, which had recently been calibrated in an ABS study of the dynamic stability of semi-submersibles (these tests were also performed at Arctec).

Cost estimates for the Spar hull were prepared by Rauma-Repola Offshore (a new group formed by Rauma-Repola to contract offshore fabrication work) based on Finnish and Korean construction. We discovered that there was no great penalty for building circular Spars, and the idea of making octagonal shapes was dropped.

Between 1990 and 1992 we continued development with the help of further studies from Chevron and other oil companies. One of the GC205 variations was a two-drilling-rig version of the Spar. Chevron continued to consider the Spar, the TLP, and the compliant tower, but no decision was made to develop the field.

By this time the Spar concept had gained some supporters in the oil companies, principally Irv Brooks at Chevron and Steve Perryman at Amoco, both members of the research establishment of those companies. We had many discussions about how to prove to the operating people and other oil companies that the Spar was a viable concept. There were questions about VIV, heave motions and such that no amount of analysis or small scale tests seemed to answer. We concluded that a large-scale offshore test was the answer, but how could we afford it? Our solution was to approach Scripps to see if we could use the FLIP ship as a large-scale Spar model. We wanted to check the VIV behavior at large Reynold's numbers. Scripps was willing, but the FLIP belonged to the Navy and could not be used unless the Navy agreed. The VIV experts at the Naval Research Lab were lukewarm to idea but generally agreed with it. However, they could not offer any financial support, and after months of negotiations, we could not get all the approval needed and had to drop the approach.

The backup plan was to initiate a joint industry program to sponsor large-scale model tests. DOT had designed a production and drilling Spar in conjunction with Reading and Bates, which was selected as the base case for the model tests. Shell agreed to join the program if its design for a small, tethered Spar with wet trees would also be tested. Amoco agreed to construct the models at its Tulsa research lab, and eight additional sponsors joined. With the recent oil company mergers, it would be difficult to get this many sponsors today.

Fig. 1-9. Large-scale model tests conducted in 1993 helped convince Oryx to select the Spar (Courtesy of Technip USA, Inc.)

The tests were conducted at the Offshore Technology Research Center (OTRC) at Texas A&M University in 1993. They exposed a wide audience of industry and academic engineers to the Spar concept and helped to validate our numerical models. The OTRC adopted the Spar as a "theme" structure for its sponsored research program, enabling many professors and graduate students to use the test data for their own research purposes.

The tests proved comprehensive. Much like the second-order heave responses during the Berkeley tests in 1988, the second-order surge/pitch responses were of particular interest in 1993. There was concern that model scale damping might produce un-conservative results. To address this concern, some experiments were conducted with a restrained model to determine the magnitude of the second-order forces. This was done by setting the mooring stiffness so that the natural period of surge was around 35 seconds. First-order (wave frequency) responses were not limited, but the slow drift responses were constrained. By imposing pairs of sinusoidal waves at different frequencies (with the difference frequency close to the unrestrained natural frequency), the drift force Quadratic Transfer Function (QTF) could be derived for a representative selection of frequency pairs. While these measurements have never been fully exploited or compared with theoretical values these tests gave the participants, in particular Oryx Energy, the confidence to pursue the Spar concept.

A Real Project

While these tests were in planning, DOT was contacted by Vic Baugus of Oryx to design a Spar for its field at Vioska Knoll Block 826. By that time, DOT was discussing a teaming arrangement with J. Ray McDermott. The Oryx work helped precipitate an agreement between DOT, Rauma-Repola, and McDermott.

A conceptual design study for Oryx was completed in early 1994. The results were sufficiently encouraging, and Oryx said they would probably go ahead with a construction contract if we could provide a competitive firm cost. Rauma Offshore (the Offshore Subsidiary of Rauma Repola), J. Ray McDermott, and DOT collaborated to provide the fixed-cost bid. The consortium was loosely structured, but the roles were clearly defined: DOT owned the Spar concept and would be responsible for conceptual design and global motions. J. Ray McDermott would handle topsides and installation, and Rauma Offshore would be responsible for the hull construction (in Finland) and mooring procurement.

In February 1995—11 years after Horton's first study of the Spar for Arco—Oryx signed the contract for the first delivery of a production Spar. It probably took more than 30 man-years to reach this point. Gaining acceptance for a new concept like this is no small task.

The Neptune Spar

As the Oryx spar took shape and negotiations were taking place between McDermott, Rauma Repola, DOT and Oryx, another company was watching the developments with heightened interest. Aker Maritime, lead by Svein Eggen, were a leader in Semi-Submersible design and construction. Svein saw the spar's potential, and on the eve of the signing of the first spar contract he was able to negotiate a deal to acquire the Finish fabrication yard and offshore business from the parent company, Rauma Repola. That became "Aker Rauma Offshore". Svein continued to champion the spar through the ensuing years and through several mergers and acquisitions.

With the contract signed, we needed to design a "real" Spar. Oryx elected to have separate contracts with J. Ray McDermott, Aker Rauma Offshore, DOT, and the other contractors. California-based DOT was responsible for the configuration and conceptual design. The detailed hull design was being carried out in Finland. Project management, topsides, and installation engineering were being done in Houston. Oryx's small technical team, lead by Vic Baugus, was working out of Dallas, Tex. There was no integrated project team as used today. The Oryx project manager, Don Vardeman, insisted that project team members be connected to the Internet, and almost all communication was by e-mail. This was before the Internet was widely used. We had used e-mail, but file sharing was another matter. DOT had a PC (286 vintage) in its Irvine office with a 20 MB hard drive, which was large for its time. We used PC Anywhere software to transfer files over phone lines to the hard drive. The DOT engineers (except Timo Lehtinen and Horton) preferred working from their homes, making electronic file sharing essential. As the Neptune project progressed, we all got faster modems and began sharing files via the Internet.

Neptune contractors met monthly as a group to resolve schedule and interface issues between contractors. Roger Glanville was the sole DOT representative at these meetings while the rest of us spent most of our time in California doing the technical work.

Fig. 1-10. The Neptune Spar (Photo provided by Rigzone.com)

As the project progressed, we had to deal with several key technical issues unique to the Spar:
1. risers and buoyancy cans;
2. upending loads;
3. keel joint and keel guide; and
4. VIV (again).

Much of the Oryx project experience has been chronicled in Offshore Technology Conference (OTC) papers. The riser system had two features unique to a Spar—buoyancy cans provided the tension, and the riser passed through a point of high bending and potential wear at the keel.

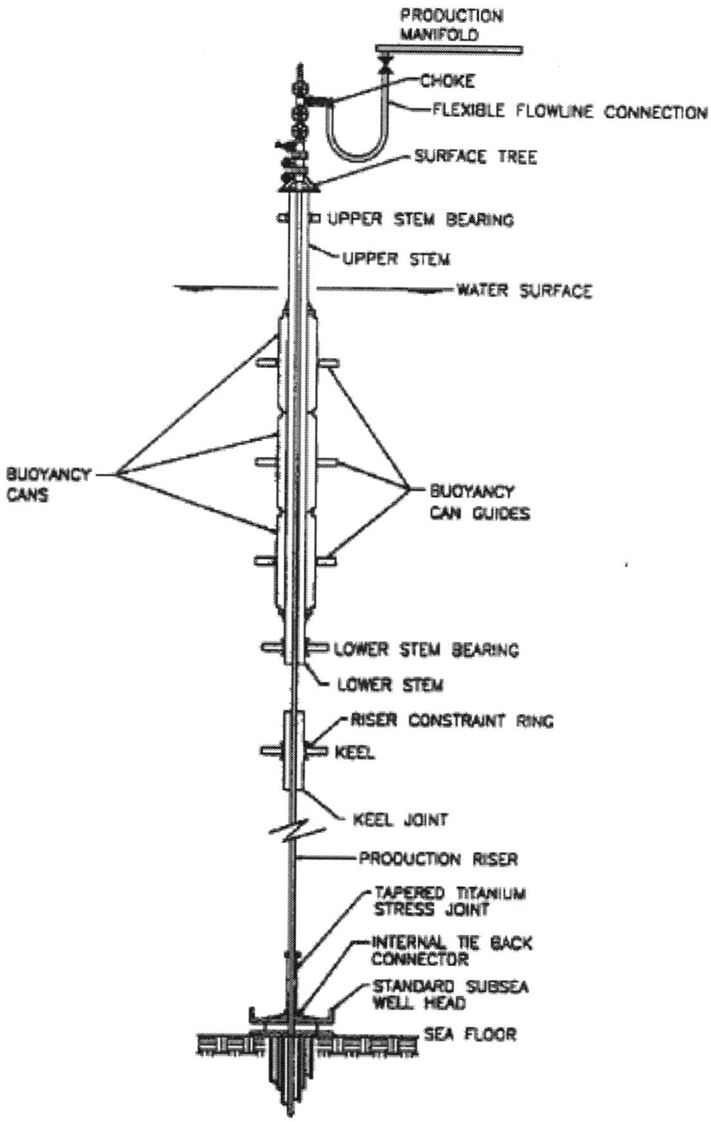

Fig. 1-11. Neptune riser system was the first to use buoyancy cans for support (Source: Prislin et al. 1999. © 1999, Society of Petroleum Engineers, Inc. Reproduced with permission of SPE. Further reproduction prohibited without permission.)

The buoyancy cans had been designed to take the full hydrostatic pressure. Buoyancy cans on subsequent Spars were designed to operate at ambient pressure. The advantage of the former is that there are no pressurized valves to leak and cause a loss of buoyancy. A disadvantage is that the cans, particularly the deeper cans, can become heavy and inefficient. The ambient air buoyancy cans may all be made the same, and they are lighter and less expensive than the hard buoyancy cans.

Fig. 1-12. Neptune buoyancy can (Source: Berner et al. 1997. © 1997, Society of Petroleum Engineers, Inc. Reproduced with permission of SPE. Further reproduction prohibited without permission.)

The keel joint, while unique, was a straightforward design. A sleeve around the main riser pipe provided wear protection and distributed the bending in the riser to two end points rather than at a single contact point. Wear data on compliant tower conductor guides were used to estimate the wear of the keel joint. This turned out to be a conservative computation.

One aspect of riser design was a surprise to those experienced in Tension Leg Platform (TLP) design. Almost all of the fatigue damage on TLPs is due to wave frequency effects and was typically computed using frequency domain methods. However, a Spar's wave frequency pitch and surge is about the same order of magnitude as the corresponding slow drift responses. The fatigue of the Spar risers, particularly near the keel, is largely a result of slow drift pitch motions. This is one reason we have always favored time domain analysis methods. We conduct fatigue calculations using rainflow-counting methods on a derived stress time history. Model tests and full-scale data have subsequently shown that damping in the real world reduces the slow drift responses significantly compared to calculations.

Another important innovation in the Spar design was the mooring system. Because of the favorable motions, the mooring legs could be designed as semi-taut members with uplift on the anchors. This greatly reduced the amount of wire and anchor chain normally associated with a spread mooring. Oryx was the first floating production system to employ this kind of mooring.

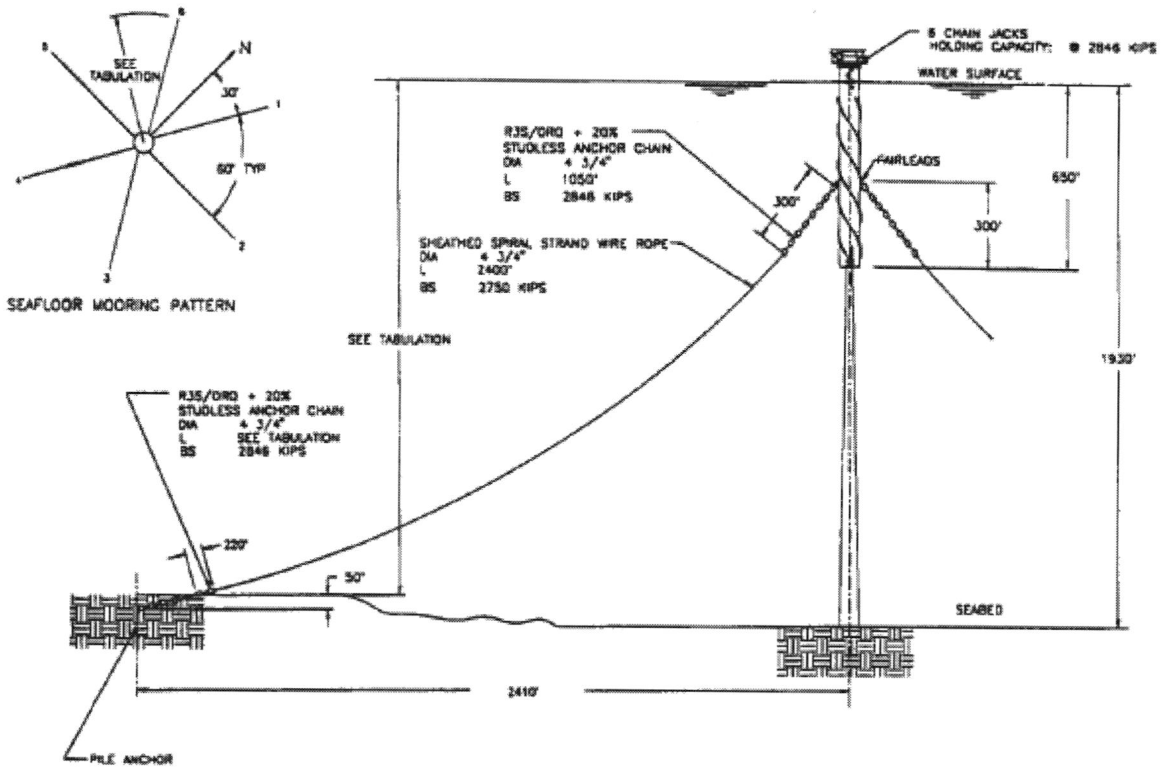

Fig. 1-13. The Neptune Spar introduced a semi-taut mooring using steel line (Source: Prislin et al. 1999. © 1999, Society of Petroleum Engineers, Inc. Reproduced with permission of SPE. Further reproduction prohibited without permission.)

Four different model tests were run as the Oryx design progressed. Three of these, conducted in Escondido, focused on the VIV behavior. Previous tests had been performed for Spars with and without helical strakes, but never with appurtenances such as pump casings and flowline conductors outside the hull. The pipes outside the hull strengthened the wake vortex, reducing the effectiveness of the helical strakes. If the current came from certain directions relative to these external pipes, the responses were almost the same as if there were no strakes. Fortunately, the design current was highly directional, and we were able to arrange the pipes to have minimal impact from the most likely current direction.

The issues with VIV have been, and continue to be, a consideration in all Spar designs. Each design had different appurtenances on the hull's exterior. Subtle changes in external fittings can have unexpected results on the VIV behavior, and we had to test almost every hull in its final configuration.

The fourth model test for the Neptune Spar was conducted at the Offshore Technology Research Center in College Station, Tex. These tests modeled the upending of the hull and its in-place responses to waves and current. Strain gauges were applied to the model's outer shell to determine bending moments during upending. This turned out to be the condition of maximum bending in the hull and an issue of some concern at this stage of development. When Ed Horton muttered the compliment to Archimedes after the successful upending, he might have also thanked the Englishman William Froude, who pioneered scale model testing in naval architecture.

After the Neptune Spar was successfully installed, we were fortunate to have a Joint Industry Project to support the installation of instruments on the Spar to measure metocean conditions and Spar responses. We had these instruments up and (mostly) running when two major hurricanes passed through the Neptune Field in September 1998. The data generally confirmed our theories and past model test experience, but one fact was quite amazing. The Spar heave motions were only a fraction of those used in our design. The pitch motions were also over predicted in our analysis.

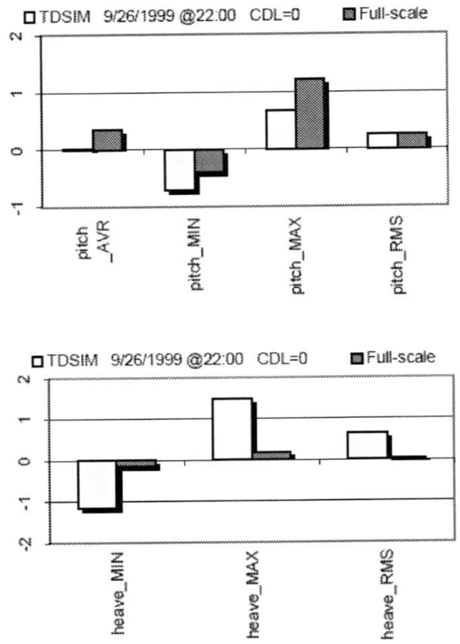

Fig. 1-14. Neptune data showed that heave and pitch motions were over-predicted (Courtesy of Technip USA, Inc.)

These effects could easily be explained. When we performed our design calculations we ignored the effect of the risers. Because the Spar is a highly tuned system and most of the heave response comes at resonance, the damping caused by the riser friction at the keel and the guides turns out to be a major factor in determining the heave response. We later learned that the drag on the mooring lines also provides a significant amount of heave damping. As a result, we were very conservative in our heave predictions. Pitch was a similar story, but for a different reason. Ignoring the risers meant that we were ignoring the pitch restoring forces from the riser tension at the keel coupled with the buoyancy can reactions at the guides in the centerwell. These restoring forces increase the effective stability (as measured by the parameter GM) of the Spar. Thus, the mean pitch angle is reduced, and the slow drift component of dynamic pitch is reduced as well. Because the pitch period remains well above the wave energy (65 seconds), there is no effect on wave frequency responses.

Interest in Spars accelerated after Oryx announced its intention to build one, and even more so after the first successful installation. Aker Rauma Offshore proved that they could build the hull on time and within budget. This was extraordinary for a novel, prototype design. It also said a lot about Oryx and its partner, CNG, and how they managed the project.

Immediately after the Oryx signing, Chevron showed renewed interest in the Spar for GC205, and we found ourselves engaged in a new conceptual design phase. This became an interesting point in the Spar evolution. DOT had been testing a new type of Spar, the Truss Spar, since 1994 and was considering proposing it to Chevron. The Truss Spar replaces the steel shell midsection with a truss structure. This saves steel weight and drag, and we thought it could potentially reduce the VIV issue. We saw Chevron's interest as an opportunity to introduce the Truss Spar design.

In February 1995 we built and tested a Truss Spar specifically sized for Chevron's field without sponsorship from Chevron. This model had no strakes because we thought the truss would provide enough damping to mitigate the VIV without them. We were again surprised on a Friday night, but this time I was the customer and was able to insist that we stay late and add strakes.

Fig. 1-15. First Truss Spar tested in February 1995 (Courtesy of Technip)

The Truss Spar includes heave plates between the bays, which have the primary benefit of trapping added mass to increase the heave period. The heave plates also add a considerable amount of damping. This allowed us to design Spars shorter than the Classic Spars. It also provided a great advantage in construction because the hulls could now be built in one piece in Finland and dry transported to the United States. The Oryx Spar (and all subsequent Classic Spars) had to be shipped in two sections and joined on the Gulf Coast. This could add up to 16 weeks to the schedule.

While the development work on the Truss Spar progressed at a fast pace in early 1995, we were competing with a semi-submersible conversion proposed by Chevron for GC 205. Chevron requested a fast-track schedule for a Spar similar in appearance to the Oryx Spar, so we dropped any efforts to promote the new Truss Spar technology for GC 205.

Fig. 1-16. Truss Spar (Courtesy of Technip USA, Inc.)

The Neptune contract was signed in February 1995. At that time Aker and J. Ray McDermott shared the licenses for the technology. They decided that the best way to promote the Spar concept was to form a separate company, called Spars International, Inc. (SII), which would serve as the marketing and project management entity for Spar projects. DOT remained a separate company, but Aker and J. Ray McDermott purchased a controlling interest. DOT continued to provide engineering support and R&D for SII.

SII's first Spar contract was the Chevron Spar with partners Exxon and Petro Fina. This one was much larger than the Neptune Spar, and it had drilling capability. The organizational structure also reflected the larger scale of a major oil company. An integrated project team was set up with Chevron and Exxon personnel staffing key positions. Front-end engineering exceeded a year, from June 1995 to July 1996. Much of the engineering effort centered on VIV. Several test programs were conducted. The delivery contract was signed in July 1996, the Spar was installed 23 months later (June 1998), and the first oil was produced 29 months after contract signing (January 1999). This schedule required steel for the hull to be ordered four months before the formal contract was signed.

The Genesis Spar used about twice the amount the steel as the Neptune Spar and took about nine months longer to build and install.

The years 1995 through 1997 were active for DOT. Ed Horton's one-man company grew to 35. The Irvine office was closed. Ed Horton and Dick Davies moved to Houston, while three of us stayed behind in California to work on new concepts. (Davies joined DOT in Irvine in 1990. He had worked with Horton at Global Marine and became the principle design engineer on the

Neptune Spar.) We were looking for new products to make the Spar obsolete. A big part of our R&D work included model tests at the offshore model basin facilities in Escondido, which was a short drive for those of us who lived in California. We performed seven separate test programs in 1995, 12 in 1996, and 10 in 1997.

Figure 1-17 presents the portfolio of new Spar concepts developed during this time. The plate Spar and Truss Spar were considered improvements over the basic Classic Spar design. The DP Spar is a drilling Spar with very low drag in currents and excellent heave motions. The TelSpar and ColSpar are both designed to be constructed in a collapsed state in a dry dock or graving dock. Decks can be preinstalled and commissioned. Once on site, the skirt sections are extended, and the platform becomes a deep-draft floater with low heave motions. The pipe Spar was designed specifically for the Caspian Sea. The shallow draft Spar shown has alternately been called the shoebox Spar, Spar box, box Spar, and oil box. It is a dry tree, large topsides, and oil storage solution for West Africa and Brazil. This concept became the DPS-2000 after SII was disbanded and the technology licenses were transferred to Aker Maritime and McDermott. The concrete Spar is a derivative of the many gravity base structures (GBS's) built in Norway in the 1970s and 1980s. It is actually much smaller than those structures but provided an interesting solution for deep-water fields in the Norwegian Sea. We conducted several studies in 1997 and 1998 for Spar use in the Norwegian Sea and the North Atlantic, but none of the projects panned out.

While DOT was carrying out this research internally, we were also involved in some interesting developments supported by oil companies. Amoco, in particular, was interested in the Truss Spar and included it in a design competition for the Marlin Field. This allowed us to advance the technology considerably. Several model tests were performed including one at OTRC and two at the Offshore Model Basin (OMB) in California. The Truss Spar indeed looked competitive, but Amoco ultimately selected a TLP for the project.

Additionally during this period, Exxon discovered oil in its Diana Field and decided to use a Spar to develop the combined Diana and Hoover Fields. The contract was awarded in 1996, making Diana the Third Spar to be built.

The Truss Spar got another boost when Amoco selected it for the King Field in 1997. DOT, SII, Aker, and J. Ray McDermott all participated in an integrated project team, which essentially completed a detailed design of this Spar, only to have the whole project dropped following BPs acquisition of Amoco in 1999.

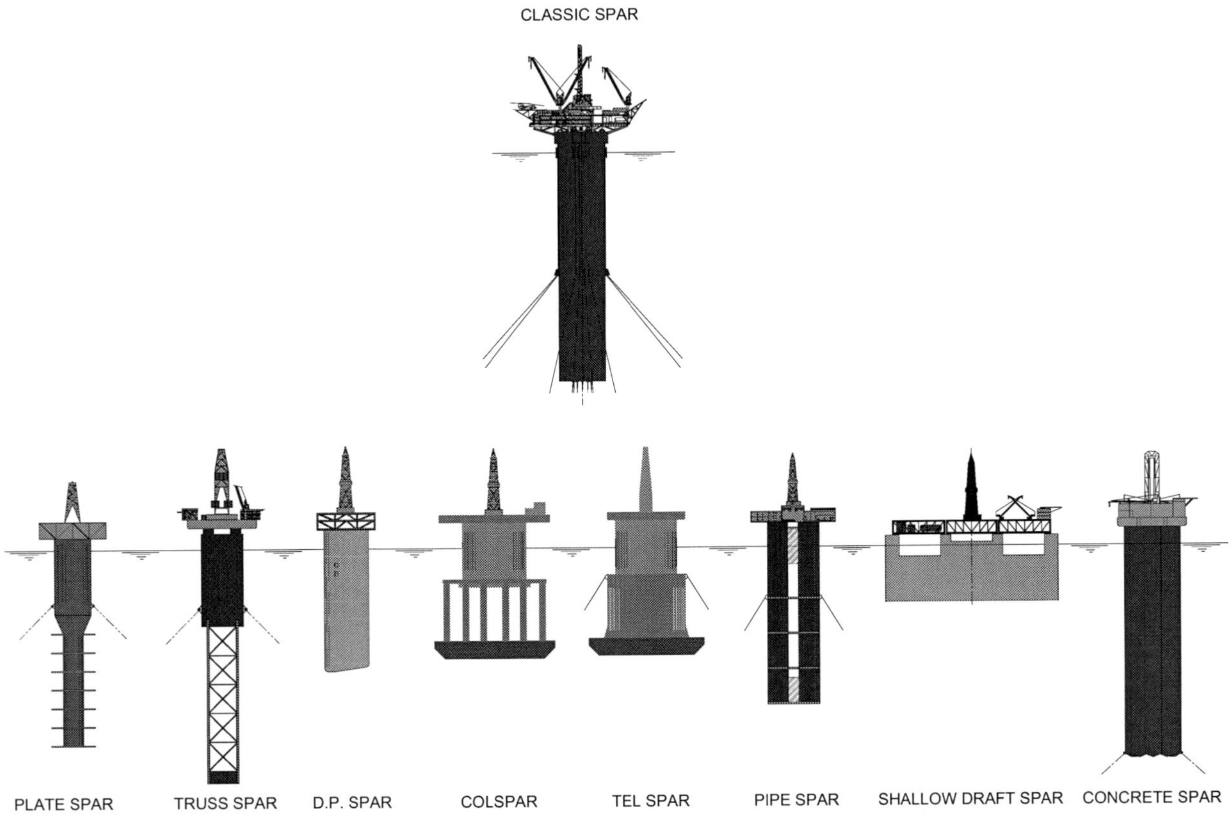

Fig. 1-17. New concepts tested by DOT 1995–98 (Courtesy of Technip USA, Inc.)

In late 1999 Aker and McDermott elected to split the SII joint venture and to pursue Spar projects on their own. This meant that they each shared equal rights to the technology that Ed Horton's DOT had developed during the previous 16 years. DOT was bought out by Aker and McDermott, and all the DOT and SII employees received two job offers. Now there were two competing Spar companies.

Prior to its breakup, Spars International had been negotiating with Kerr McGee (Kerr McGee had bought out Oryx by this time, and was later bought itself by Anadarko Petroleum) for the design, construction, and installation of the first Truss Spar, Boomvang. Boomvang had a twin spar, Nansen, which ultimately became the first truss spar contract to be signed by partners Kerr McGee and Ocean Energy. After some delay in receiving partners' approval, Kerr McGee finally signed a contract for the Boomvang Truss Spar in June 2000. Because of the history of these negotiations, the EPCI contract was awarded to Spars International even though the company no longer existed except on paper. Subcontracts were let to McDermott for the topsides and installation and to Aker Maritime for the hull and moorings, following the previous SII business model. A third Truss Spar was ordered shortly after the Nansen order from Aker Maritime by Vastar Resources, who had been studying the Truss Spar for a number of years. The Vastar spar, Horn Mountain, was to become the first of three spars that BP built, as they bought Vastar shortly after the Horn Mountain spar contract was awarded. Thus, after a hiatus of nearly three years since the last Spar contract (Exxon Hoover/Diana) and the breakup of SII, three Truss Spar contracts were signed during the course of a few months. These orders were followed in quick succession by more Truss Spar orders: Aker received orders from Kerr McGee for Gunnison and from BP for Holstein and Mad Dog. McDermott's new company SparTec received orders from Murphy for Medusa and later for Front Runner and Dominion Devil's Tower Spar.

After the Breakup

After the SII breakup up, Aker was able to recruit most of the DOT engineers. Aker split its engineering department and put Paul Stanton, a long-time Exxon Production Research employee who joined DOT in 1998, in charge of a new Riser engineering department. I was put in charge of the new product technology department with a nice R&D budget to continue the new product development tradition of DOT. Lyle Finn, another EPR veteran, became chief engineer for the Aker engineering organization. The rest of the DOT team was broken up and integrated into the various engineering groups within Aker. Unfortunately, or fortunately, Ed Horton was restricted from working for either Aker or McDermott for a period of one year from the SII breakup. Thus, he was left alone to exercise is inventive mind.

Aker became busy staffing up to manage the Spar projects, especially those contracted by BP. The first innovation under the new regime was the development of the compliant guide for the Horn Mountain Spar buoyancy cans.

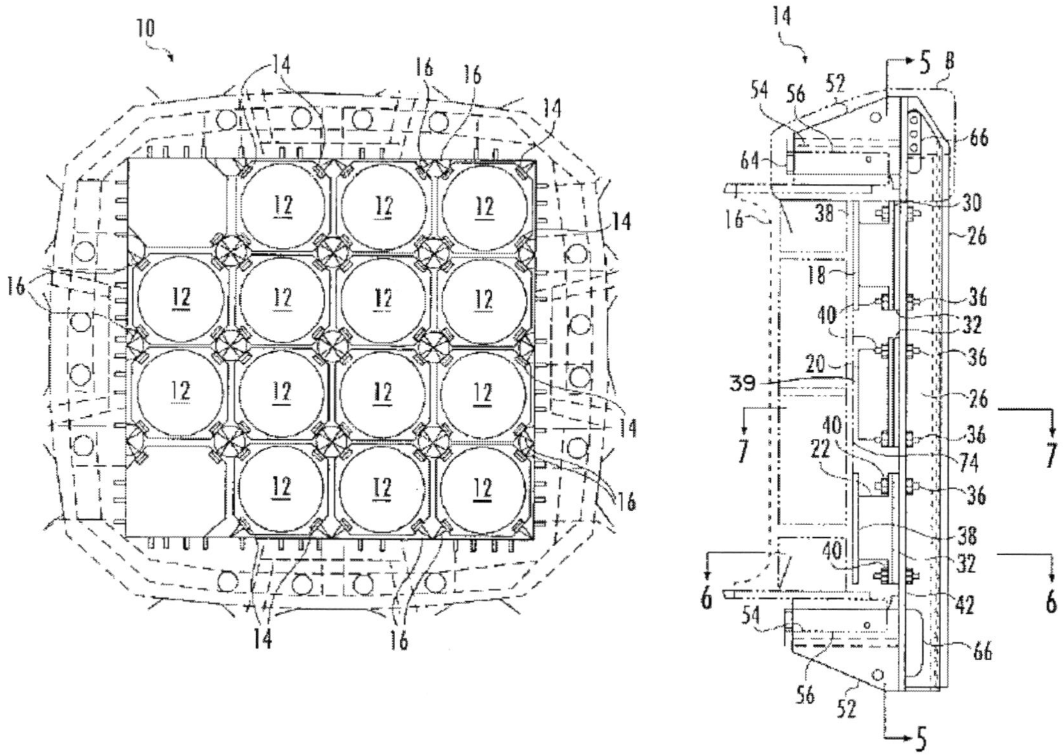

Fig. 1-18. Compliant guides were developed for the Horn Mountain Spar (Source: U.S. Patent 6,679,331)

The compliant guides provided stiff, lateral constraint of the buoyancy cans, which prevented the buoyancy cans from impacting a rigid guide surface when the Spar surges. All subsequent Spars have used this type of guide for the buoyancy cans, and the Genesis Spar has been retrofitted.

Pushing Floatover

Other R&D efforts in 2000 and 2001 focused on one of the most vexing issues of Spar projects—deck installation. The Spar had to be towed to deep water and upended before the deck could be installed. Small decks could be installed with a derrick barge, but larger decks required multiple lifts and long and expensive offshore hookup times. The decks of semi-submersibles and TLPs could be installed and commissioned quayside. We wanted to be able to do the same thing; however, the commissioned deck would have to be floated over the Spar.

Our new owner, Aker Maritime, was quite familiar with large deck floatovers, having done numerous floatover deck installations on gravity-based structures and semi-submersibles. These were mostly done in the calm waters of the fjords in Norway near Stavanger or Haugesund. Offshore floatovers had been performed on fixed platforms, but none had been done on a floating platform.

There are three main challenges to performing a floatover on a Spar offshore:

1. The deck needs to be supported on two barges in a catamaran configuration. There was concern about whether the loads caused by this method of transportation would significantly affect the deck structure.
2. Relative motions between the deck and Spar during mating require the use of shock cells or other energy absorbing devices, and the critical phase of mating should be done as fast as possible.
3. The Spar needs to be deballasted rapidly to complete the operation and separation of the barges.

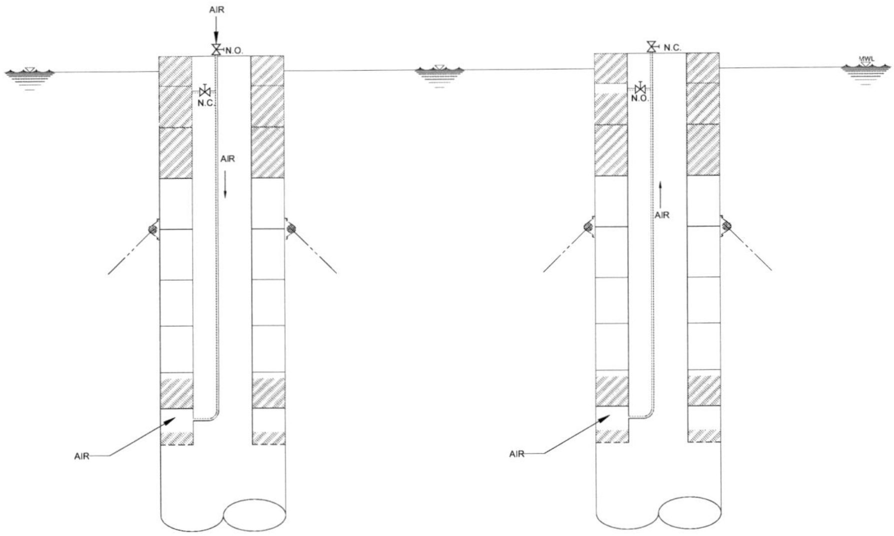

Fig. 1-19. Stored-air deck-lift system allows rapid load transfer during deck mating (Courtesy of Technip USA, Inc.)

We had previously conducted extensive research using small waterplane hulls for Spar floatovers in the swell conditions of West Africa. We had also invented a unique method for deballasting the Spar—the stored-air deck lift system (SADL). This method uses compressed air stored in the variable ballast tanks (200 ft. below the water level) to blow ballast from tanks at the waterline. While not quite as efficient as using hydraulic cylinders, it would allow transfer of sufficient weight to secure a 15,000 ton deck on the Spar in less than 10 minutes.

Fig. 1-20. Model tests proved the effectiveness of small waterplane pontoons for deck floatover in West African swell (Courtesy of Technip USA, Inc.)

Our goal in 2000 was to qualify the floatover method for five Spars that BP was proposing for the Green Canyon area of the Gulf of Mexico. BP had decided to develop the large Holstein, Mad Dog, Atlantis, and Crazy Horse (now Thunder Horse) fields using standardized technology. Semi-submersibles were our competition, and we wanted to surmount the handicap of having to perform multiple derrick lifts. We brought Aker's floatover experts from Norway to Houston for a presentation to BP and other operators in March of 2000. We enlisted John Montague from Offshore Kinematics (now Dockwise) to talk about the success his company had had with large shock cells for floatover. This meeting was followed with more technical work and a major model test in the fall of 2000 demonstrating the floatover of a 35,000 st deck onto a Spar. Altogether more than $1 million was spent trying to qualify this method of deck installation in the Gulf of Mexico. Although McDermott had successfully installed the Auger TLP deck this way in 1994, there were still no champions for this method among the major US operators.

Fig. 1-21. Extensive development of floatover technology for large decks in the Gulf of Mexico failed to win supporters from the operators in the early 2000s (Courtesy of Technip USA, Inc.)

In the end, two Spars were built (Holstein and Mad Dog) using conventional deck lift methods.

In 2001 Aker's deepwater group in Houston was taken over by Coflexip-Stena Offshore (CSO). The name became CSO-Aker for a year, although Aker was no longer in the picture. We kept pushing the floatover method, however, and it was bid as the method of deck installation for the ExxonMobil Kizomba and Agbami Spars in West Africa, but CSO-Aker failed to land either of these contracts.

In 2002 Technip took over CSO, so we became Technip-Coflexip and later Technip. Once again we had a parent company that was a world leader in floatover deck mating, this time mainly on West African jackets. We kept pushing for acceptance of this technology for Spars. Finally in 2005, Murphy Oil Company agreed with our proposal to use the floatover method for the Kikeh Spar in Malaysia. The Kikeh deck was relatively small as decks go, about 4500 ton, but it was just too big for the largest derrick barges in Southeast Asia to lift in one piece. This led to the world's first offshore floatover deck installation on a Spar in the South China Sea off Sabah in East Malaysia in November of 2006.

Fig. 1-22. Kikeh was the first Spar floatover, November 5, 2006 (photo by the author)

The Cell Spar

Ed Horton's year of isolation came to an end in 2001, and he teamed with CSO-Aker to form a new company, Deepwater Technology, Inc. (jointly owned by Horton and CSO-Aker). He brought new ideas with him.

One of these ideas was the Cell Spar. This consisted of a number of tubes, each about 20 ft. in diameter, which were assembled to form a ring of tubes to provide buoyancy and support for a deck. The motivation was to reduce fabrication costs and schedule time. The Classic and Truss Spars (except for the trusses) were constructed using ship building practices. Plate panels were strengthened with stiffeners and supported on girders. Construction required a lot of welding on a panel line and the assembly of many pieces. Many intersections of stiffeners with girders and decks needed to be detailed. The tubes of the Cell Spar could be rolled and ring stiffened. They didn't need longitudinal stiffeners, eliminating much of the welding, and this fabrication method was familiar to Gulf of Mexico fabrication yards. The hull could be constructed in the Gulf, thus avoiding the long transportation from Finland. Horton also proposed setting and grouting the deck to the hull, rather than welding it. This provided many advantages in time and fabrication.

Horton had made a "Home Depot" model of the Cell Spar in his garage. One day Don Vardeman from Kerr McGee visited and saw it. Kerr McGee had been struggling to find a way to develop their Red Hawk Field. The company was looking at tie backs to another operator's platform or a stand alone or hub development. Vardeman liked the Cell Spar, saw a possible fit with Red Hawk, and encouraged Horton to continue development. After all, Vardeman and his team had been behind the first Spar (Neptune) and the first Truss Spars (Bomvang and Nansen). Why not the first Cell Spar?

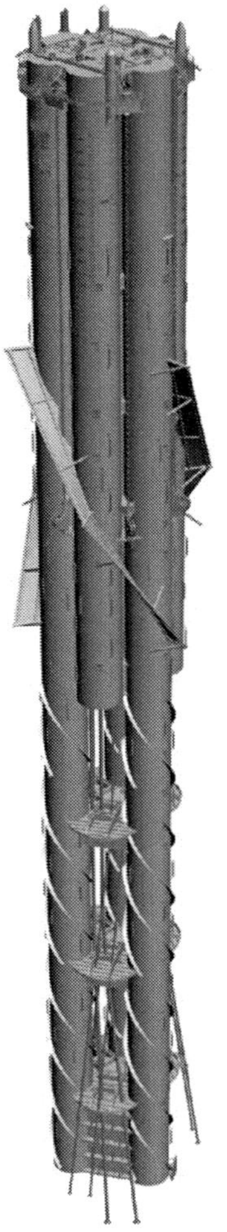

Fig. 1-23. Red Hawk Cell Spar (Courtesy of Technip USA, Inc.)

There were still questions about the Cell Spar. How strong would the VIV be, and could it be suppressed with helical strakes? Remember that back in 1990 DOT tried to use an octagonal Spar. The idea was abandoned because of VIV, which could not be mitigated by strakes. Some thought that closing the space between the cells (the closed cell version) would be required to allow the strakes to be effective. Additionally, it was unclear how the lower stem would behave. The first design had a single lower cell extending below the hard tank sections. We wondered whether we would need helical strakes for this section. We also were unsure of its motion behavior and whether we needed to add heave plates.

A series of model tests were conducted in Escondido in January 2002 with a yellow (eight-cell) model and a red (six-cell model). The yellow Spar was tested with and without closure between the cells, and with and without strakes. Different strake designs were tested for both the upper and lower hull.

Fig. 1-24. The first Cell Spar had a single lower column (Courtesy of Technip USA, Inc.)

The tests proved that VIV could be handled with helical strakes and that the Cell Spar motions were at least as good as Truss Spars.

Shortly after we completed the model tests, Kerr McGee asked Technip to prepare a bid for a Red Hawk Cell Spar. This was to be an EPCI contract where Technip would take full responsibility to design, build, and install the hull and deck. It was unusual to go directly from R&D to an EPCI contract. Normally there was a front end engineering design phase to work out all the kinks in the design. Nevertheless, Technip prepared a bid and submitted it to Kerr McGee.

Everyone at Technip was depressed when Kerr McGee announced that it had selected ABB to build its single-column floater (SCF) for Red Hawk. Erling Storaune, my boss and the head of the Floater Product group at Technip, held a commiseration party at his house the night the decision was made (Erling worked for Svein Eggen and was part of Aker Maritime at the time it purchased Rauma Offshore in 1995. He served as Vice President of Spars International. He was a key guy in recruiting the DOT Engineers after the split of Spars International in 1999).The only happy person in the room was Horton. He was nervous about the readiness of the Cell Spar and was actually relieved. There was much work to do, and Horton was as enthusiastic about the Cell Spar as ever and wanted to press on.

We were just at the stage of looking at the global strength of the design and had decided that the single lower tube would not be adequate. We also had new ideas for the strake design and decided to return to the model basin one more time.

It turned out that Kerr McGee had awarded the Red Hawk contract to ABB contingent on the successful performance of additional model tests. ABB had proposed a shortened version of their SCF design, which would allow for quayside deck installation. We returned to Escondido for the second Cell Spar tests in June 2002. Luis Bensimon of Kerr McGee was in San Diego on a surfing vacation and was able to drop by to witness the tests.

Shortly afterwards, Kerr McGee contacted us again and wanted us to rebid the Cell Spar for Red Hawk. Apparently, the shortened version of the SCF had some issues that arose as a result of the model tests, putting the Cell Spar back in contention.

Fig. 1-25. Red Hawk Cell Spar under tow (Courtesy of Technip USA, Inc.)

Hurricane Alley

The years 2004 and 2005 were two of the worst for hurricanes in the Gulf of Mexico on record. Hurricane Ivan passed directly over the Devils Tower, Horn Mountain, and Neptune Spars in 2004. Aside from damage to the rigging on the Horn Mountain workover rig, loss of the completion rig on the Devils Tower Spar, and some superficial damage to the Neptune topsides, the Spar hull and moorings were undamaged. In 2005, Hurricanes Katrina and Rita passed through the central gulf areas as category 5 storms. None of the Spars were seriously affected, although one mini-TLP was completely lost to Hurricane Rita.

The Future of the Spar

The Spar is now an established concept for deepwater developments, particularly in the Gulf of Mexico where it accounts for about 40 percent of the deepwater production. The Spar and the TLP remain the only proven deep water dry tree solutions. The choice of a Spar, semi-submersible, or TLP is always a difficult decision—and one that cannot always be made based on numbers. There are important soft issues like the personal preferences of key operator staff, timing, and the availability of certain human and physical resources. The early 2000s were a boom time for Spars and other deepwater concepts. However, the fastest growing segment in deepwater is subsea. As this technology matures, especially subsea processing, the demand for floaters can be expected to reduce.

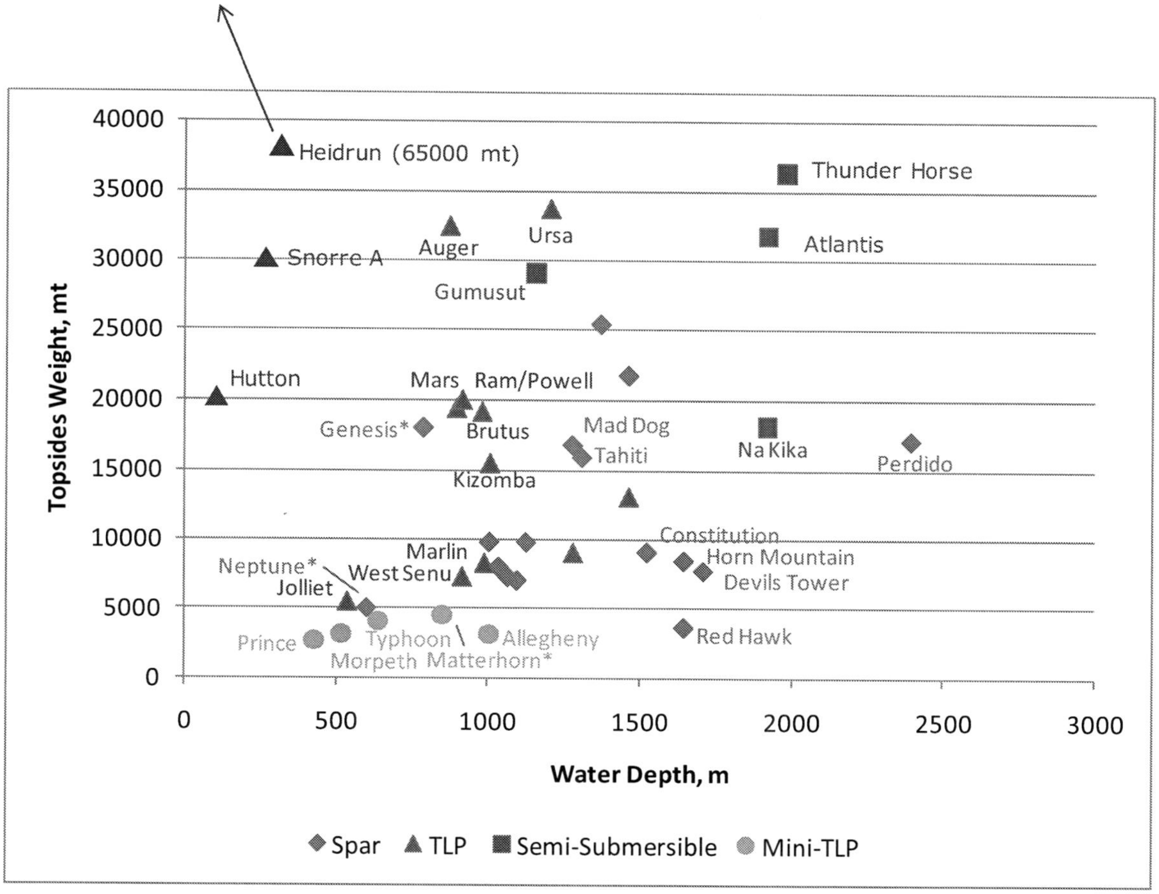

Fig. 1-26. After only 12 years, Spars have a major position in deepwater development

The Kikeh Spar was the first to be installed outside the Gulf of Mexico. Many in the industry believe that the future for Spars is in Southeast Asia, the North Sea, Brazil, and elsewhere. The Cell Spar represents a potential solution for marginal fields, especially where in-country fabrication is desirable.

Many contractors are designing concepts to compete with the Spar, always trying to reduce the cost and development schedule. Nevertheless, Spars have proven their merit and will probably remain on the short list of deepwater concepts for many years.

Acknowledgment

The author would like to thank his previous employer, Technip, for allowing permission to publish this paper. He would especially like to thank Steven Barras of Technip for his comments and suggestions on improving the final manuscript.

Appendix A: Spars in Place or Under Contract as of August 2007

Project	Year	Operator	Type	Location	Water Depth, ft	Contractor	Oil, bpd	Gas, mmscfd	TTRs	Drilling	Topsides, st	Hull, st	Dia., ft	Length, ft	Mooring
Neptune	1996	Oryx Energy	Classic	GOM	1960	J Ray McDermott/Aker Maritime/Deep Oil Technology	25000	30	16	workover	5500	13000	72	705	6 - 4.75" Chain/Strand
Genesis	1998	Chevron	Classic	GOM	2598	Spars International	55000	72	20	full drilling	17000	28700	122	705	14 - 5" Chain/Strand
Hoover/Diana	2000	ExxonMobil	Classic	GOM AC 25/26	4800	Aker Maritime (SII License)	100000	325	6	full drilling	24000	35000	122	705	12 - 5.6" Chain/Strand
Medusa	2002	Murphy	Truss	GOM MC 582	2224	SparTec	40000	110	6	workover	9000	13000	94	586	10
Nansen	2002	Kerr McGee	Truss	GOM	3678	Spars International	40000	200	9	workover	8800	12000	90	543	9 - 4.9" Chain/Strand
Boomvang	2002	Kerr McGee	Truss	GOM	3451	Aker Maritime(Technip)	40000	200	5	workover	8800	12000	90	543	9 - 4.9" Chain/Strand
Horn Mountain	2002	BP	Truss	GOM	5423	Aker Maritime (Technip)	65000	68	14	workover	9000	14600	106	585	9 - 5" Chain/Strand
Gunnison	2003	Kerr McGee	Truss	GOM	3150	Aker Maritime (Technip)	40000	200	9	workover	10700	13800	98	549	6 - 4.9" Chain/Strand
Front Runner	2004	Murphy	Truss	GOM	3330	SparTec	60000	110	8	workover	11000	14000	94	586	9
Holstein	2004	BP	Truss	GOM GC 645	4344	Technip	110000	90	15	full drilling	27000	35550	149	746	16 - 5.9" Chain/Strand
Mad Dog	2004	BP	Truss	GOM	4419	Technip	80000	40	16	full drilling	18500		128	555	11 - 5.8" C/10.1" P
Red Hawk	2004	Kerr McGee	Cell	GOM	5299	Technip	15	300	wet tree		4400	7000	64	560	6 - 4.5" C/8.5" P
Devils Tower	2004	Dominion	Truss	GOM MC 773	5610	SparTec	60000	60	8	workover			94	586	
Constitution	2006	Kerr McGee	Truss	GOM GC 680	4970	Technip	70000	200	6	workover	10770	14800	98	554	
Kikeh DTU	2007	Murphy	Truss	Malaysia	3937	Technip			10	Tender Assist			106	466	10
Tahiti	2007	Chevron	Truss	GOM GC 640	4000	Technip	125000	70	wet tree						
Perdido	2008	Shell	Truss	GOM AC 857	7817	Technip	130000	150							

Appendix B: Milestones in Spar Development

1962	FLIP Built
1976	Brent Spar Installed
1984	Deep Oil Technology, Inc. Formed Ed Horton Invents Spar Production System
1984	Spar Concept presented to ARCO Earle & Wright Review Concept
1987	JIP Formed to Study Spar Motions (UC BerkeleyTowingTank) Spar Patent Issued
1988	JIP Formed to Study TBT Motions (Arctec Offshore Model Basin)
1989	Spar evaluated for Chevron R&D Department for GC205
1987	JIP Formed to Study Spar Motions (UC BerkeleyTowingTank) Spar Patent Issued
1988	JIP Formed to Study TBT Motions (Arctec Offshore Model Basin)
1989	Spar evaluated for Chevron R&D Department for GC205
1990	VIV Tests for Chevron (UC Berkely Towing Tank)
1990	DOT and Rauma Repola enter marketing agreement: keeps DOT alive!
1992	Failed at attempt to use FLIP as a large scale test bed!
1993	Multi-Company JIP Model Tests at OTRC
1994	Oryx asked for design and "real" costs for VK 826 (Neptune) field DOT and Rauma Offshore form Joint Venture with McDermott
1994	DOT Tests New "JACKET Spar" (Truss Spar)
1995	Oryx signs contract for Neptune Aker buys Rauma Offshore
1995	Chevron picks Spar for GC205 Aker and McDermott form Spars International
1995–1998	DOT performs over 30 R&D test programs. 16 Truss Spar tests!
1996	Chevron signs contract for Genesis Spar
1996	Exxon executes license for Diana-Hoover DDCV
1996	Preliminary Truss Spar Design for Amoco MARLIN Field in competition with TLP (TLP Selected)
1996	Oryx Installs Neptune Spar on time within budget, Aker and J. Ray McDermott form Spars International
1997	Truss Spar Selected for King Field (not built)
1998	50 year Hurricane passes over Neptune Spar, full scale measurements of global responses recorded
1998	Genesis Spar Installed
1999	Diana DDCV Installed
1999	Aker and McDermott Separate
2000	Three Truss Spar Contracts Awarded
2001	Compliant Guide Developed for the Horn Mountain Risers Aker Maritime Deepwater group sold to Coflexip-Stena Offshore, becomes CSO-Aker

2002	Cell Spar Tests, Red Hawk Contract Awarded, Technip buys CSO, becomes Technip-Coflexip
2002	Four Truss Spars installed
2004	Hurricane Ivan directly over Horn Mountain and Neptune Spars as Cat 3 Storm
2005	Hurricane Katrina and Rita Strikes Gulf. Rita's eye passes over Red Hawk.
2005	Chevron Tahiti award, first wet tree Truss Spar.
2006	Shell Perdido award, deepest floating production unit, first with Direct Vertical Access risers Keppel Fels and J. Ray McDermott form Floatec, acquire ABB TLP patents Kikeh Floatover Installation: first floatovedr on a Spar and first Spar installed outside of the Gulf of Mexico.

References

Berner, P.; Baugus, V.; Gendron, K.; Young, R. (1997). "Neptune Project: Production Riser System Design and Installation." Proceedings, Offshore Technology Conference.

Fisher, F. H. and F. N. Spiess. (1963). FLIP-floating Instrument Platform, *J. Acoustic Soc. of America,* 35, 1633-16544

Halkyard, J., Davies, R., Paulling, R., and Glanville, R. (1991) "The Tension Buoyant Tower: A Design for Deep Water", Proceedings, Offshore Technology Conference, Houston.

Prislin, I.; Halkyard, J.; DeBord, Jr., F.; Collins, I.J.; Lewis, J.M. (1999). "Full Scale Measurements of the Oryx Neptune Production Spar Platform Performance," Proceedings, Offshore Technology Conference.

Chapter 2: Taut-Leg Mooring System and Anchoring for Spars

By Shukai Wu. Ph.D., MODEC, Principal Engineer

Introduction

Background and Operating Principles

Types of Mooring System

Mooring systems are used for the stationkeeping of floating offshore structures, such as platforms for drilling and production of oil offshore. Figure 2-1 shows four types of offshore floating platforms using different types of mooring systems for stationkeeping.

A graphic illustration of the different mooring systems is shown in Figure 2-2. A tendon mooring system consists of a set of nominally vertically tensioned tendons used for mooring TLPs. A spread mooring system consists of multiple mooring lines (or legs) that are spread around the perimeter of the platform.[1]

Fig. 2-1. Examples of floating offshore platforms

[1] The turret-mooring used for the FPSO shown is a spread mooring type.

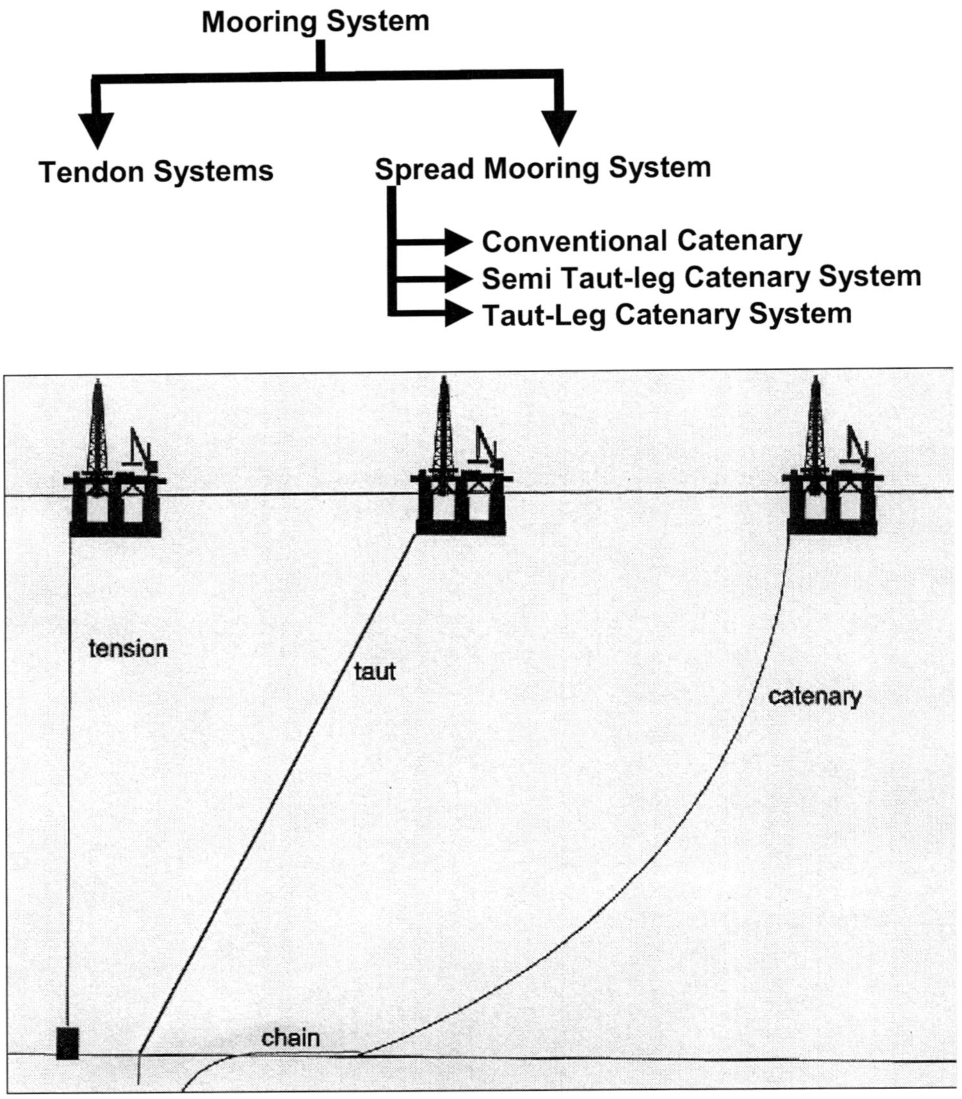

Fig. 2-2. The types of mooring systems

Spread mooring systems are generally catenary. The suspended vertical line profile of each of the mooring lines is of catenary shape, defined by the well-known catenary equations. For ease of discussion, they are further divided into three categories:

1. *Conventional catenary mooring systems* typically use drag-embedment anchors with little anchor uplift loads (zero uplift angle for conventional anchors and a maximum of 10° anchor uplift angle for Bruce and Stevpris high-holding capacity anchors). Examples include those typically used on spread-moored mobile offshore drilling platforms.
2. *Semi taut-leg catenary mooring systems* (or semi taut-leg systems for short) impose no uplift on the anchors in the nominal pretensioned condition and a maximum uplift angle of more

than 10° under full design top tension. Usually, resistance to anchor uplift requires the use anchors different than those for conventional catenary mooring systems.

3. *Taut-leg catenary mooring systems* (or taut-leg system for short), which due to the taut catenary profile of the lines, have no grounded line lying on the seabed in the nominal pretensioned condition. As a result, the mooring line anchor in a taut-leg system experiences an uplift lift angle (at the nominal burial point of the line) under pretension. Furthermore, the mooring line's uplift lift angle at the burial point can significantly exceed the 10° limit for high-holding capacity drag-embedment anchors.

Because of the dry tree requirements, which require a tight watch circle of the vessel offset, taut or semi-taut leg mooring system designs have been used on existing spar platforms. This chapter deals with taut-leg systems for spars.

Working Principles

In a spar mooring system, the lower end of each leg is anchored to the seabed. It is then led through a fairlead to the top of the spar hull, pretensioned through a tensioning device (usually chain jack), and held in place at the stopper (Fig. 2-3). The ABS Guide for Building and Classing Floating Production Installations (ABS June 2000) provides a more general definition of the mooring system applicable to spar as well as other types of floating offshore platforms, including mooring lines, connectors and hardware, winches, piles, anchors, and thrusters.

The static vertical profile of an individual mooring line is defined by the well known catenary equation. Because its anchor point is fixed, a mooring line becomes tauter as the top end at the fairlead is moved away from the anchor, resulting in an increase in the line tension. Conversely, the line becomes slacker and the line tension decreases as the top end of the line moves towards the anchor (Fig. 2-4).

A mooring system utilizes the horizontal force component of the mooring lines to provide the restoring forces to keep a spar on station. The mooring system set-up is defined with reference to the zero static equilibrium position of the spar, that is, the equilibrium position of the spar subject to no environmental loads except the forces of the catenary mooring system. Any movement of the spar off station induced by environmental or other loads causes increased tensions and a tauter (less steep) catenary configuration in the wayside (on the opposite side of the offset) mooring lines and reduced tensions and a slacker (more steep) catenary profile in the leeside (downward of the offset) mooring lines. The net effect is a restoring force about opposite of the spar offset.

In addition, change of the catenary profile of the mooring lines in water subject to dynamic excitation at the spar's fairleads has a damping effect on the movement of the mooring lines and the spar mainly due to the hydrodynamic interaction between the lines and seawater. The modeling and analysis of the dynamic effects are dealt with in another chapter.

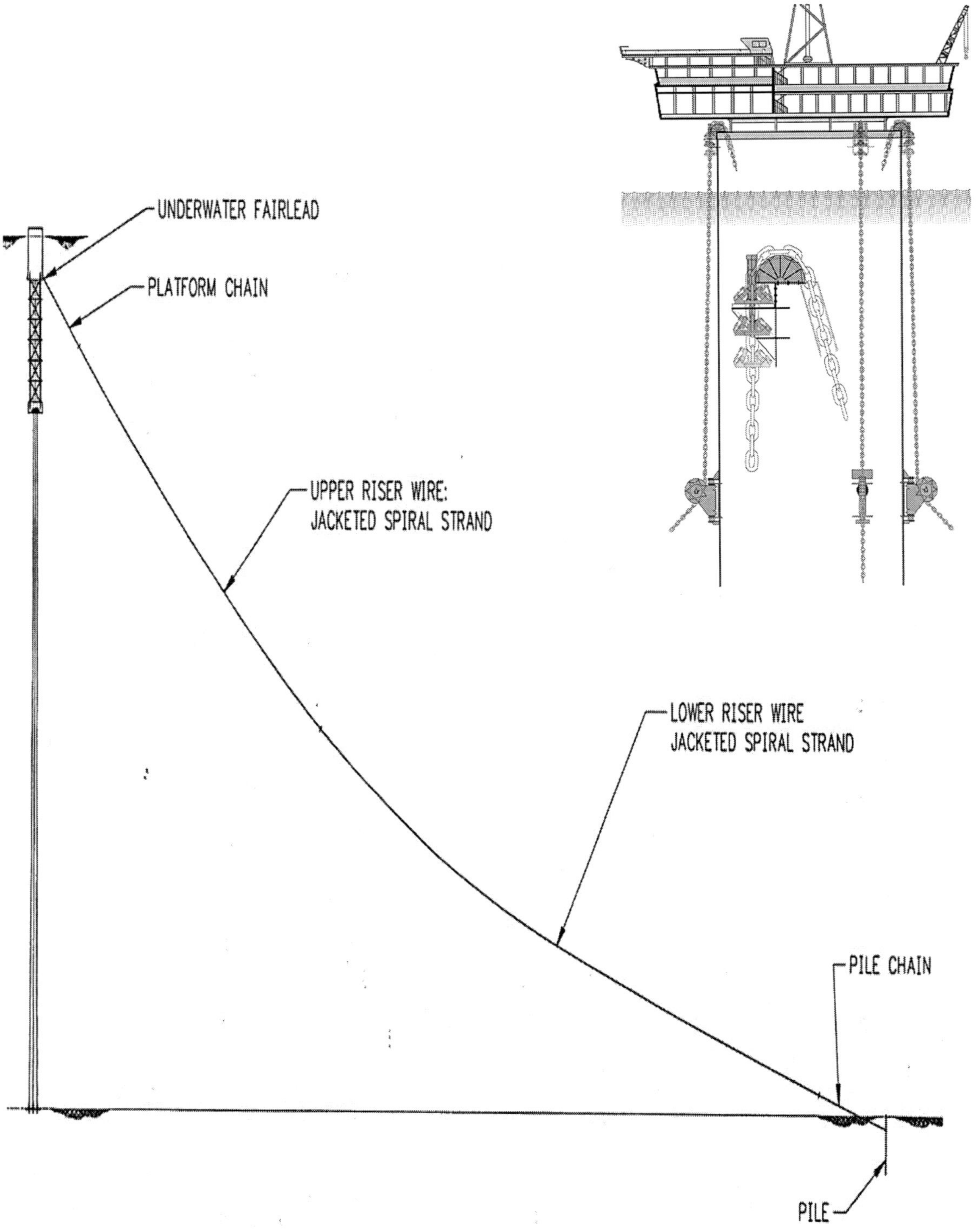

Fig. 2-3. Makeup of a spar mooring system

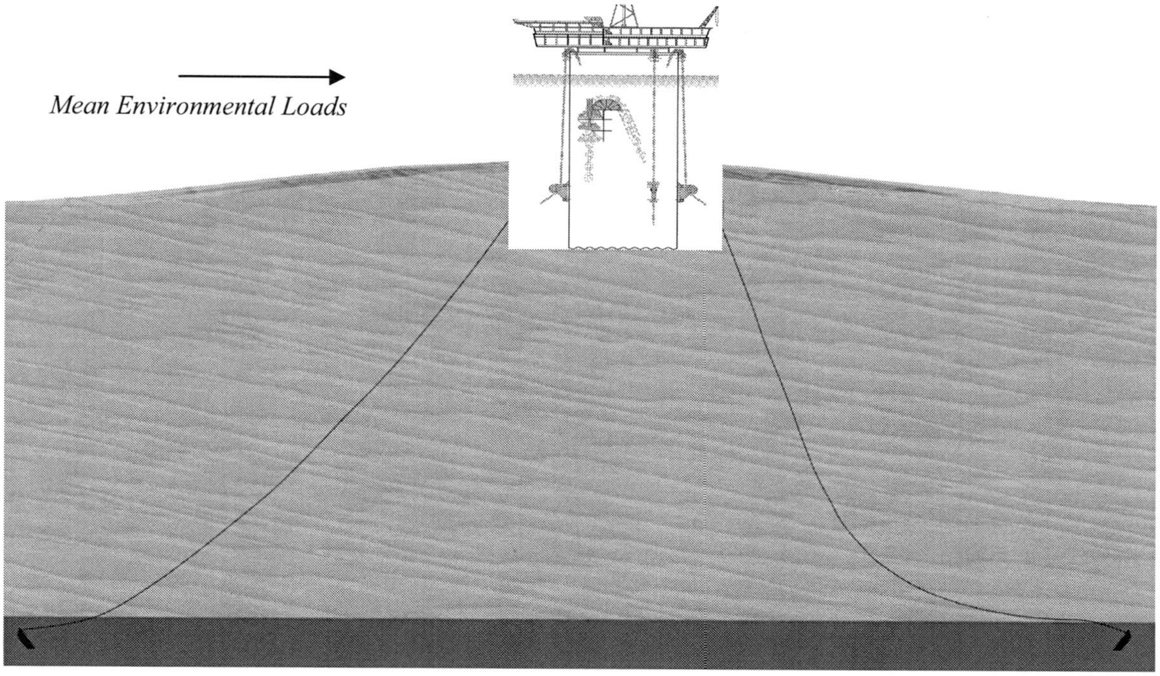

Figure 2.4. Working principle of a catenary mooring line

Mooring Designs on Existing Spars

Table 2-1 shows details of the mooring system design for all of the spar platforms installed in the Gulf of Mexico as well as those waiting to be installed and those being designed.

Mooring Criteria

Key Design Issues

When the spar is subjected to the design extreme environments, it experiences a large offset from its static equilibrium position. Consequently, some of the lines in its mooring system become highly loaded, whereas others become slacker and more lightly loaded.

The following occurs within the most loaded line:

- Maximum line tension reaches a significant portion of its breaking strength.
- Maximum line tension on the anchor reaches a significant portion of the anchor holding capacity.
- The maximum anchor uplift angle significantly exceeds the tolerance for conventional drag-embedment anchors.

For the least loaded line, on the other hand, significant grounding may occur, that is, a significant length of the line at the lower end is grounded on the seabed. As a result, the dynamic natural grounding associated with the line movement induces abrasion and added wear of the mooring line.

Table 2-1. Mooring System Design Details

	KMG	ChevronTexaco	Exxon	KMG	KMG	BP	KMG
	Neptune	Genesis	Hoover/Diana	Bomvang	Nassen	Horn Mountain	Gunnison
Spar Type	Classic	Classic	Classic	Truss	Truss	Truss	Truss
Location	VK 826	GC205	AC 25	EB 643	EB 602	MC127	GB 668
Water Depth	1930	2590	4800	3300	3700	5400	3100
Size							
Diameter	72	122	122	90	90	106	98
Hard Tank Depth	220	240	295	188	188	176	
Draft	650	650	650	493	493	549	
Rig System	Workover (1,000HP)	Full Drilling (25,000ft)	Full Drilling	Workover (1000hp)	Workover (1000hp)	Workover	Workover (Nabors API 1500hp boottrap rig)
Design Criteria	100-Yr w/o Rig	100-yr w/ Rig	100-yr w/ Rig	100-Yr w/o Rig	100-Yr w/o Rig	100-Yr w/ Rig	100-yr w/ Rig
	50-Yr w/ Rig			50-Yr w/ Rig	50-Yr w/ Rig		
Mooring System Type	Steel (chain-wire-chain)	Steel	Steel	Steel	Steel	Steel	Steel
Mooring System Details	Pullover Drilling			Pullover Drilling (300ft)	Pullover Drilling (300ft)	Pullover Drilling (310ft)	Pullover Drilling (400ft-both sides)
No. of Lines	6	14	12	9	10	9	9
Configuration	6 x 1 Semi-Taut	14 x 1 Semi-Taut	4x3 Taut	3x3 Taut	3x3 Taut	3x3 Taut	3x3 Taut
Platform Chain							
Grade	R3S (ORQ+20%)	R4 Studless	R4 Studless	R4 Studless	R4 Studless	R4 Studless	R4 Studless
Size	4-3/4"	5-1/4"	6"	5-1/2	5-1/2	5-3/4"	5.59"
Length (o/b of frld)	300	750	250	6x230'+3x430	6x230'+3x431	250-450'	400
Procurement Length	1,050		800	700ft - 900ft	700ft - 900ft	850ft - 1000ft	850-950
MBL	2,846		4246 kips				
Manufacturor		Vicinay	Vicinay	Vicinay	Vicinay	Vicinay	Vicinay
Wire							
Type	JSSW Rope	JSSW Rope	JSSW Rope	JSSW Rope	JSSW Rope	JSSW Rope	JSSW Rope
Size	4.79	38108	5-5/8"	5		5"	4.92"
Length	2,400	2000	6640	3900' - 5050'	3900' - 5200'		4300-5100
Procurement Length	2,400		6640	3900' - 5050'	3900' - 5200'		4300-5100
MBL	2,750		3320 kips				3,420 kips
Manufacturor			Trefileurope	Bridon	Bridon	Bridon	Bridon
Ground Chain							w/ ball-grab connector
Grade	R3S (ORQ+20%)	R4 Studless	R4 Studless	R4 Studless	R4 Studless	R4 Studless	
Size	4-3/4"	5-1/4"	6"	5-1/2	5-1/2	5-3/4"	5.59"
Length	190-230	250	225	250	251	250	250
Procurement Length	190-230		225	250	251	250	250
MBL	2,846		4246 kips				
Anchors							
Type	Driven Piles	Driven Piles	Suction Piles	Driven Piles	Driven Piles	Suction Piles	Driven Piles
Size (OD x length)	84"x180ft	96" (84" head) x 235'	21ftx105ft	84" x 250ft	84" x 250ft	18ftx90-93ft	84" x 220ft
Penetration	180	220ft	100ft	220ft	220ft	90ft	205ft
Weight							
Fairleads							
Type	Rotary	Rotary	Rotary	Rotary	Rotary	Rotary	Rotary
Manufacter		Pusnes	Pusnes	Pusnes	Pusnes	Pusnes	Pusnes
Tensioning Device							
Type			Linear Chain Jack	RamWinch	RamWinch	RamWinch	RamWinch
Manufacter			Bardex	Pusnes	Pusnes	Pusnes	Pusnes
Installation Vessel	DB50	DB50	Saipem 7000	DB50/Balder	DB50	Balder	Q4000

Table 2-1. Mooring System Design Details *(continued)*

	Murphy	Murphy	Dominian	BP	BP	KMG
	Medusa	Front Runner	Devil's Tower	Holstein	Mad Dog	Red Hawk
Spar Type	Truss	Truss	Truss	Truss	Truss	Cell
Location	MC 582	GC 338	MC 773	GC 645	GC 826	GB877
Water Depth	2223	3330	5610	4300	4420	5300
Size						
Diameter		94	94	149	128	
Hard Tank Depth				236		64 (6x20')
Draft		500		691		
Rig System	Workover Rig (750HP)		Workover Rig	Full Drilling	Full Drilling	N/A
Design Criteria				100-yr w/ Rig	100-yr w/ Rig	100-Yr
Mooring System Type	Steel	Steel	Steel	Steel	Polyester	Polyester
Mooring System Details						
No. of Lines	10	9	10	16	11	6
Configuration				4x4 Taut	11x1 Semi-Taut	
Platform Chain						
Grade		R4 Studless		R4 Studless		
Size		5-3/4"		6-3/4" (171mm)		
Length (o/b of frld)				250		
Procurement Length				900		
MBL						
Manufacturor				Vicinay	Vicinay	
Wire						
Type	JSSW Rope	JSSW Rope		JSSW Rope	Marlow Super-line	Marlow Super-line
Size				5-7/8"		
Length				6600'		
Procurement Length				6600'		
MBL			3750 kips	4700 kips	4,260	
Manufacturor			Bridon	Bridon	Marlow	
Ground Chain						w/ ball-grab connector
Grade		R4 Studless		R4 Studless		
Size		5-3/4"		6-3/4" (171mm)		
Length				250-650		
Procurement Length				4x250'+ 12x650'		
MBL				4700 kips		
Anchors						
Type	Driven Piles			Suction Piles		
Size (OD x length)	84"x259			21'x102ft		
Penetration	247ft			99ft		
Weight	115-127					
Fairleads						
Type	Rotary	Rotary	Rotary	Rotary	Rotary	
Manufacter	Pusnes	Pusnes	Pusnes	Rolls-Royce	Pusnes	
Tensioning Device						
Type	Roller Ram Winch	Roller Ram Winch	Roller Ram Winch	Chain Jack	Ram Winch	
Manufacter	Pusnes	Pusnes	Pusnes	Rolls-Royce	Pusnes	

In addition, there is always the possibility of a mooring line being damaged due to over-loading, among other reasons. The mooring system must be able to withstand storm environments that cause the line damage without experiencing progressive failure.

The dynamic motions of the spar subject to waves result in changes in mooring line tensions. In addition, the spar may experience VIM when subjected to currents, which also produces changes in mooring line tension. These changes in turn induce mooring line fatigue, and their long-term effects need to be considered in permanent mooring system installations. API RP 2SK contains a loose definition of permanent versus temporary mooring system.

Apart from any wearing and abrasion of the mooring lines caused by direct seabed contact in the dip zone or with fairleads and stoppers, mooring lines are subject to corrosion. The spar mooring system design must incorporate adequate allowance for wearing, abrasion, and corrosion.

Design Criteria

Mooring system design criteria typically envelope the following aspects of system performance requirements:

1. stationkeeping capability requirements—maximum watch circle both for design extreme and design operating environments;
2. mooring line tension safety factors;
3. anchor holding capacity safety factors;
4. mooring line (and other component) fatigue damage safety factors;
5. mooring line grounding; and
6. winch tensioning capacity for specified operating condition(s) and other project-specific requirements.

For steel mooring lines, a minimum corrosion allowance is required in addition to localized wearing and abrasion allowances, if any. For polyester rope, the minimum line-tension/cycle specification also limits localized fiber compression and damage.

For connector design, a combination of corrosion allowance and corrosion protection may be incorporated depending on the type (and sometimes size) of the connector in question. The issue is usually addressed in detail engineering as part of the hardware suppliers' component design.

Strength and fatigue life requirements for mooring connectors and foundations for on-vessel mooring components are usually treated the same as other critical hull structures. The foundations are usually designed for the mooring line's full breaking strength.

Metocean conditions associated with the various performance requirements discussed become an integral part of the design criteria together with the soil conditions used for designing the anchors.

Codes and Standards

Principal design codes include the following:

Codes/Standards	Principal Application
API RP 2SK	Mooring/stationkeeping system design—steel
API RP 2SM	Synthetic mooring system design
API Spec 9A	Wire rope specifications
API Spec 2F	Mooring chain specifications
DnV POSMOOR	Hydrodynamic coefficients of mooring lines
API RP 2A	Driven pile design, load, resistance factors
API RP 2T	Driven pile holding capacities safety factors
AISC Manual of Steel Construction	Anchor structural design
ABS Guide—Floating Production Installation	Mooring system design
ABS Guide—Certification of Mooring Chain	Chain specifications
DoE HSE Offshore Installations	Anchor and connector fatigue
DnV Certification Notes No. 2.6 (1995)	Chain specification

Design Drivers and Principal Considerations

It is the job of a mooring engineer to develop a cost-effective mooring system design optimized for the intended purpose. An optimized mooring system design meets the design criteria, performance requirements, and installation constraints. It also minimizes the installed cost of the mooring system. Often, however, minimization of the mooring system cost cannot and should not be done in isolation; the engineer needs to account for the performance and cost impact of the mooring system design on other spar systems, particularly the riser systems.

This section discusses key design considerations in developing a taut-leg spar mooring system.

Water Depth

Water depth is a key design parameter driving the design of a spar mooring (and riser) system. In very deepwater, large mooring lines must usually be used with significantly high pretensions to meet the design watch circle requirements. The line size and maximum design tensions drive the anchor size, the on-vessel mooring components (fairlead, chain jack, and stopper), and their foundation designs. In addition, the total vertical mooring line tension carried by the spar increases. Consequently, water depth significantly affects the direct procurement cost of the mooring components, the system installation cost, and the costs associated with increased vertical mooring line tension.

Spar Type

Depending on the type of spar—classic spar, truss spar, or cell spar—the size, environment loads, and the resulting static and dynamic responses differ, which in turn affects the mooring system design. A classic spar attracts significantly more environmental loads, particularly current drag, than a truss or cell spar of the same diameter. Therefore, it usually requires a heavier mooring system (larger size or more lines). A classic spar also has significantly more dynamic mass, being several times that of a truss spar for the same payload.

Spar Functions

Spars can be designed to fulfill different functions, such as drilling, completion, and workover of the wells and/or production. Typically, for spars designed to support so-called dry trees, drilling (full drilling or limited drilling, such as sidetracking) and/or completion and workover functions have been incorporated into the design in addition to production.

Not only do the functional requirements and operating philosophy significantly affect the payload a spar must be designed to carry and, therefore, the size of the hull and deck, they also affect the selection of design environment conditions and their impacts on the mooring system design.

For example, the Diana-Hoover, Genesis, Holstein, and Mad Dog spars were all designed with full drilling capabilities. The Horn Mountain spar was designed with completion and sidetracking capability, and the three existing spars operated by Kerr-McGee were all designed with completion only. For the Horn Mountain spar (as for the Kerr-McGee spars), post-installation drilling of any new wells will be done using a separate MODU by pulling the spar over sideways. The significant size differences between these spars are attributable to the differing functional requirements as well as differences in production capacity. In addition, the Kerrr-McGree spars are designed for completion operations in the off-hurricane season. This operating philosophy justified the use of a 100-year return storm environment without the rig (less windage) and a much reduced 50-year return storm environment with rig onboard.

Environment Conditions

The severity of environmental conditions is a key design driver. For example, a spar would require a heavier mooring system in the typical Gulf of Mexico environment than conditions expected offshore West Africa, other factors being equal.

Typical environmental conditions to be considered for spar mooring system design include:

- design extreme environment;
- reduced extreme environment (for hull flooded case);
- design maximum manning environment beyond which voluntary evacuation of platform personnel would be initiated;
- design maximum operating environment—production limit;
- design full drilling and/or completion environment;
- design limited drilling environment;
- storm-ready environment—hull only;
- storm-ready environment—pre-commissioning phase; and
- environment conditions for pullover drilling (see "pullover drilling" for further discussion).

For the last two conditions in particular, the anchor holding capacity is normally critical because the soil is not fully consolidated as discussed in the section on contract models. For the storm-ready environment—hull only, in which a minimal number of mooring lines are assumed to be connected, the line tensions may be critical even when a reduced environment is applied for this temporary condition. Additional environmental conditions are related to the design of other spar systems, such as the top-tensioned risers, but they do not usually affect mooring system design.

Other environmental conditions may also be considered, often depending on client- or project-specific requirements. For example, some of the past spar projects have included a design survival environment corresponding to the hurricane environment for a 1,000-year return period. For the Horn Mountain spar mooring system design, a 100-year sudden Gulf of Mexico environment condition was also considered. In this condition, the workover system was assumed to have full system payload with design setback when applying a hurricane condition originating from the Gulf and having a return period of 100 years. The premise is that hurricanes originating from the gulf, though less severe than the full population hurricanes, can set in without warning. The adequacy of the mooring system design, which may have a different set of pretensions under normal workover or well completion conditions than those adjusted to be hurricane-ready, has to be checked.

An environmental condition is meant to envelope all the important parameters of the metocean environment affecting the design of a spar mooring system, such as coincidence, directionality, and persistence, as well as the magnitude of wind, waves, and current.

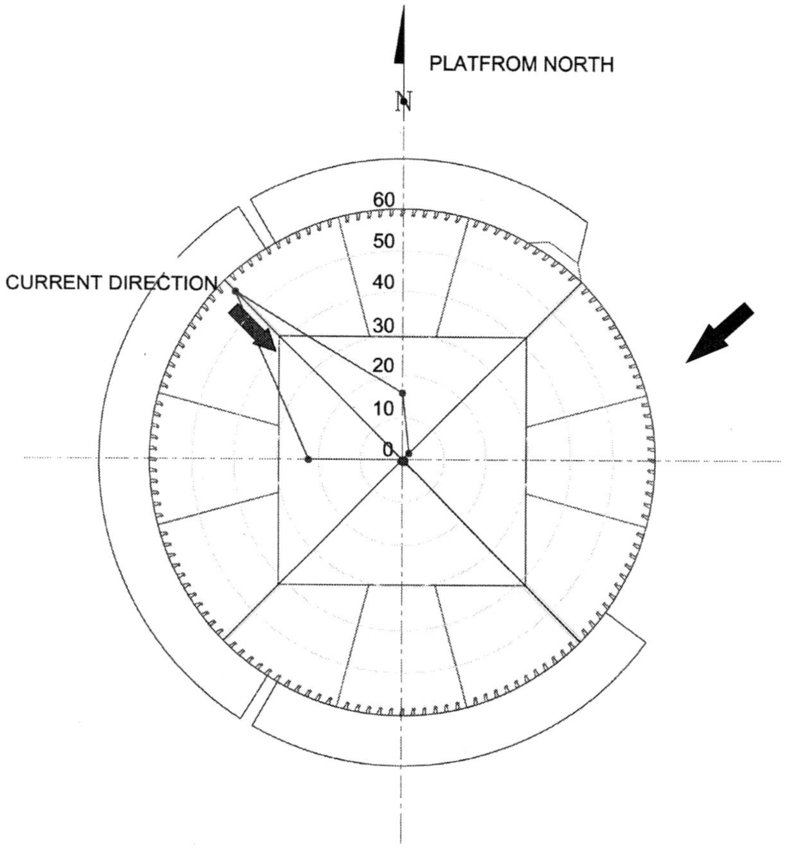

Fig. 2-5. Location of VIV strakes relative to predominant current

As an example of the impact of environment directionality on spar mooring system design, Figure 2-5 shows the direction of the current relative to the VIV strakes, which affects the location of the strake cut-off needed for dry transporting the spar hull. The cut-off location in turn affects the azimuthal location of the fairleads. It also affects the hull VIV performance, as shown in Figure 2-6. The VIV performance in turn affects the mooring system design.

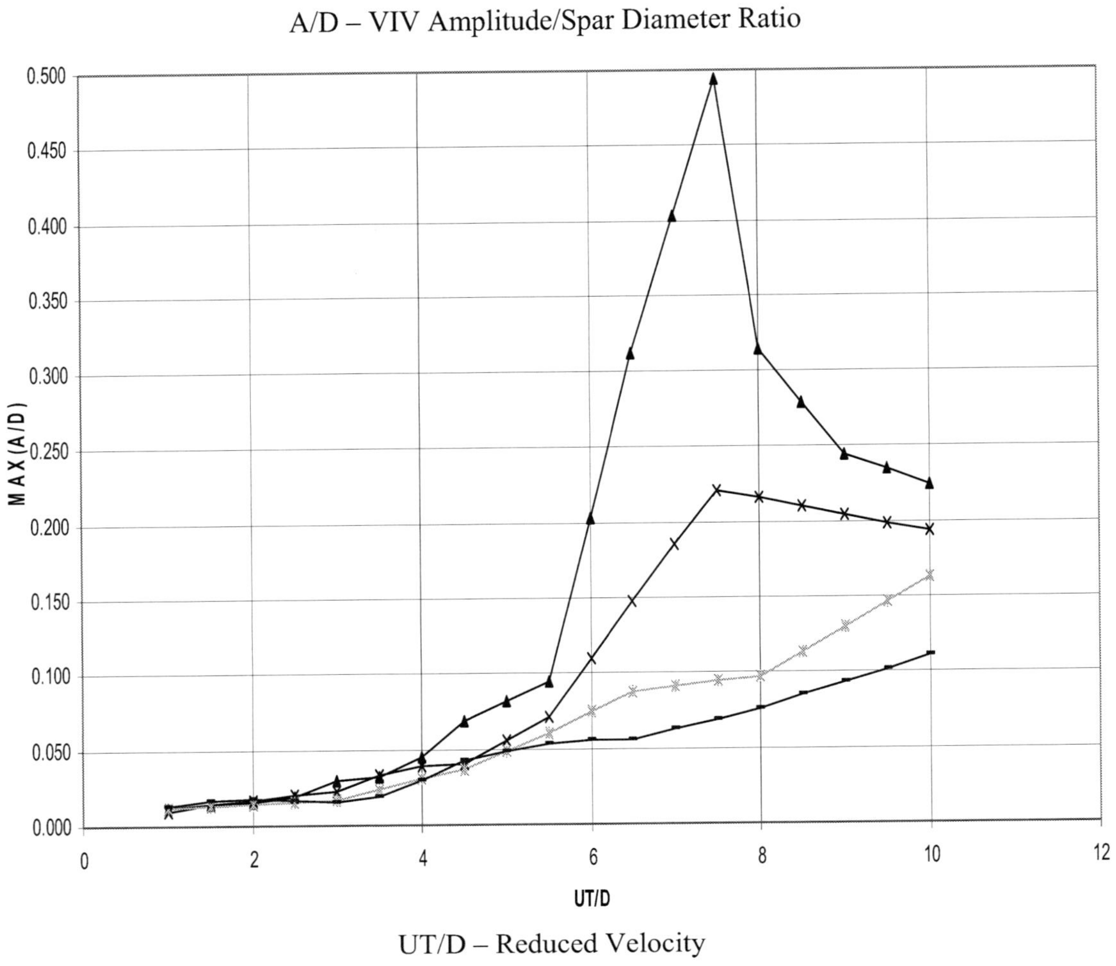

Fig. 2-6. Example spar VIV curves from model tests

Gulf of Mexico spar design project experience indicates that a spar mooring system design is usually governed by the design extreme environment condition. This is not necessarily the case for offshore West Africa, where the difference in the severity between the 100-year design extreme environment and the lesser environmental conditions, such as 10-year design maximum operating environment, is very small.

The design safety factor required depends on the environment (see "Mooring System Design" for more information on code requirements). The required minimum mooring line tension safety factors are provided in API RP 2SK. These factors are given separately for the system intact and one-line damaged (one-line broken) scenarios, either of which may govern depending mainly on the number of mooring lines in the system. The system intact scenario tends to be more critical for systems with a large number of mooring lines (typically 12 and more). For mooring system designs with fewer lines, the one-line damaged scenario tends to be more critical.

System Design Life

The design service life of a mooring system affects the fatigue life required, system configuration, mooring line size, and hardware selection.

For permanent mooring systems, the minimum required safety factor for component fatigue life is three if the component is field inspectable and replaceable, and 10 otherwise. This in turn affects the hardware selection. For example, studless chain design is generally favored for use instead of studlink chain due to improved fatigue performance as well as weight efficiency (higher strength/weight ratio). Field experience with studlink chains indicates frequent problems with loosening of the stud on the chain links due to high stress concentration and/or fatigue, severely affecting the strength and fatigue of the chain. There is limited public-domain fatigue test data on medium to large diameter studless chains; however, chain manufacturers have provided their proprietary data on various projects for studless mooring chains. These tend to show a much improved fatigue curve compared with the API RP 2SK curve for ORQ chain.

Depending on the type of design, wire ropes have significantly varying service life expectancy. This is further discussed under "Hardware Selection."

For taut-leg systems, the system fatigue design life may drive the system design, as discovered on the Amoco King/King's Peak spar design project. In such cases, the mooring line size and/or system configuration may need to be adjusted to meet the design fatigue life requirement.

Existing spar platform installations have been designed for a field operating life of 20 years or longer, and their mooring systems have been designed to be in operation for the duration of the field design life without replacement due to strength, fatigue, corrosion, and abrasion. Accordingly, jacketed spiral strand wire rope has been used to ensure adequate design service life.

The design service life also affects the corrosion and abrasion allowance, which in turn directly affects the mooring line size. The code recommended corrosion allowance is 0.4mm/year (2SK/POSMOOR). Field experience indicates that a higher allowance of 0.6 to 0.8mm/year is needed particularly in the splash zone. An additional 0.25-in. total abrasion allowance may be made for the platform chain.

Corrosion and abrasion allowances are usually made only for the chain sections. Allowance is not normally provided for jacketed spiral strand wire ropes. For mooring line connectors with corrosion protection (usually in the form of anodes), corrosion allowance is usually spared provided appropriate selection of the fatigue curve is made.

All corrosion and abrasion allowances reduce the effective size and, therefore, the MBL of the mooring chains for design. For example, for a 6-in. R4 studless chain with a catalog MBL of 4540 kips, the effective chain size with a 0.8mm/year combined corrosion and abrasion allowance for a 20-year design life (total allowance of 0.8 mm/year x 20 years = 16 mm) would be 5.37 in. (6.0 in. to 0.63 in.) with an adjusted design MBL of 3800 kips. This would be the design chain breaking strength to calculate the safety factors for mooring line strength. For the mooring line fatigue calculation, the breaking strength of a 5.37-in. ORQ chain would have to be used to calculate the line tension ratios (unless the chain manufacturer can provide his own test data acceptable to the classification society for the project).

Under certain circumstances, a mooring system designed to be field replaceable during its design service life could be justified.

Well Systems

It is well known that the requirement for dry-tree or wet-tree completion substantially affects the design of a spar as well as its mooring system. First, a much larger watch circle can usually be tolerated for spar designs with all wet-tree completions than for those with dry-tree completions. Second, the riser system, particularly the top-tensioned riser system, attracts hydrodynamic loads. These are passed on in part to the spar hull and are, therefore, resisted by the mooring system. The riser system also contributes to the restoring stiffness of the spar. Third, the motions of the riser system and the spar are coupled dynamically, the degree of which depends on a number of factors. The design of the mooring system is even more closely coupled with that of the riser system for top-tensioned risers that are directly supported by the spar hull (the so-called spar-supported vertical riser or SSVR system), as in the Holstein spar design.

Well system details affect the mooring system design for spars with dry-tree completion. For example, the size (length and diameter) of the buoyancy cans supporting the top-tensioned risers located in the spar centerwell affects the mooring system design. On truss spars with top-tensioned risers designed for very deepwater, the buoyancy can length is usually limited by the desire to avoid penetrating or interfering with (at the down-most riser stroke position) the uppermost heave plate. This length limit may lead to an increase in the diameter of the buoyancy can and, consequently, the diameter of the spar. In addition, the exposed portion of buoyancy can below the bottom of the hard tank attacks hydrodynamic loads to be resisted by the spar's mooring system.

The well pattern also affects the mooring system design. Typically, the spacing between adjacent wells at the seabed is about 1 percent of the water depth to avoid interference, whereas the well spacing at the spar deck level is smaller and is driven by three dimensions, jumper hang-off/clashing, and/or buoyancy can OD. For drilling or completing individual wells, the spar needs to be positioned directly above the well by adjusting the mooring system (see Figure 2-7). Therefore, an active mooring system is required.

The seabed well pattern should be laid out to impose the least constraints on the drilling/completion sequence of the wells to minimize interferences and spar offsets/mooring system adjustments for well completion.

Maximum spar offset is one of the principal factors contributing to riser stroke. In addition, maximum spar offset in combination with maximum pitch/roll angle drives the maximum riser tension/bending at the keel and bending moment on the stress joint. Therefore, a tight spar offset watch circle together with a reduced maximum pitch angle is preferred in designing top-tensioned risers. For spars supporting dry-tree completion risers, 5 percent water depth intact and 7 percent water depth damaged would be reasonable design targets for maximum spar offsets in the mooring system design. For drilling and completion operations, a much tighter watch circle would be needed, which could govern the mooring system design.

For spars designed with pullover capability for post-installation drilling, the well pattern on the seabed directly affects the well access and, therefore, the drilling sequence, pullover distance, and direction. For further discussion see "Pullover Drilling."

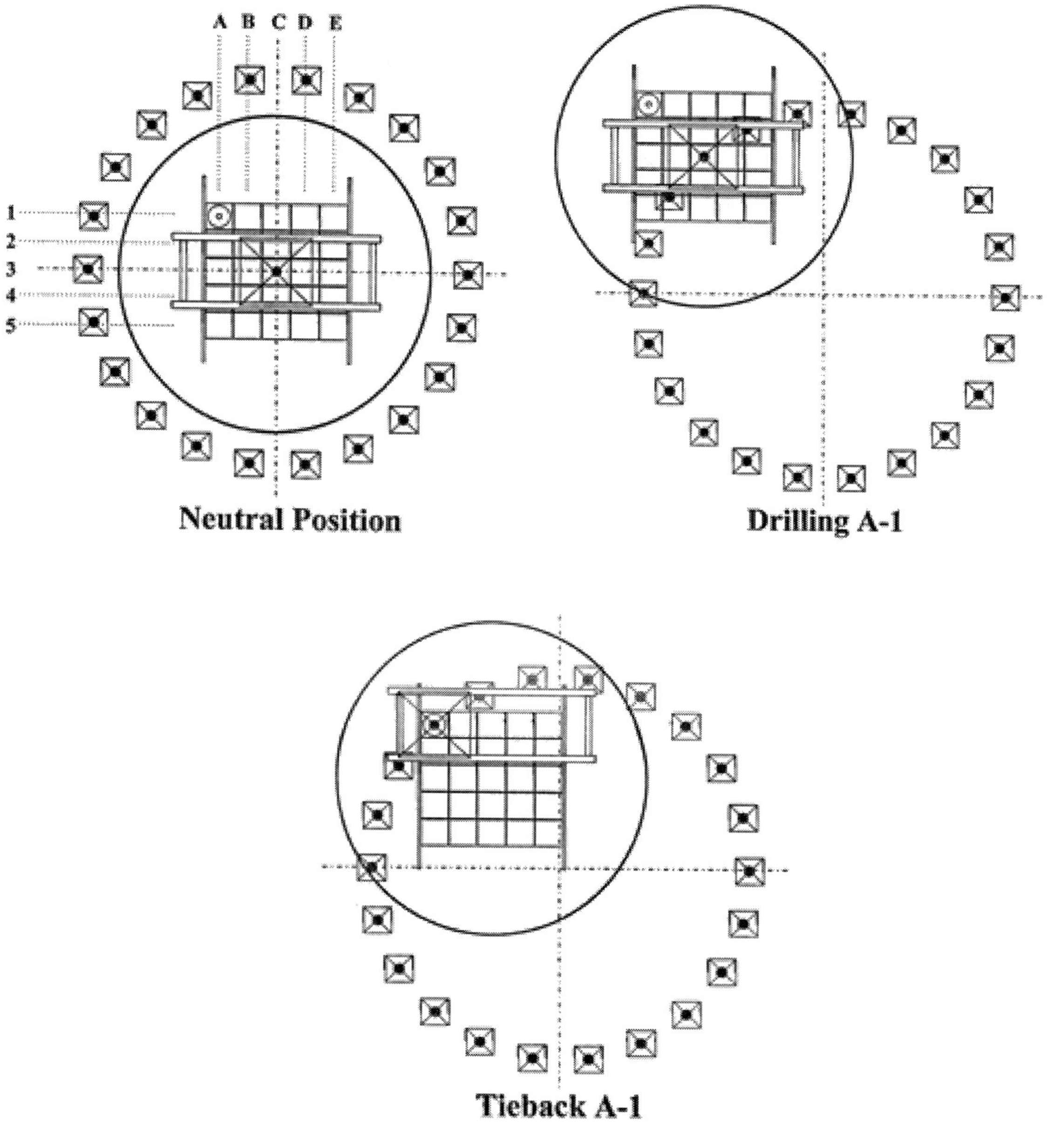

Fig. 2-7. The need for an active mooring system for positioning spar for drilling and completion operations

Pullover Drilling

Although pullover drilling can be grouped broadly into one of the spar functionalities discussed, it deserves special attention because four existing spars have incorporated this requirement in their mooring system designs—the Neptune, Nansen, Bomvang, and Horn Mountain spars. In particular, pullover drilling was used for the Neptune spar using the spread-moored MODU, Ocean Lexington. (See Fig. 2-8 for an illustration of pullover drilling.)

Important parameters in defining pullover drilling include

- spar nominal pullover direction(s) and distance(s), and
- maximum design pullover environments.

One or more pullover directions may be designed (for example, two directions directly opposite each other) depending, among other things, on the well access requirement associated with the drilling and completion programs.

The nominal pullover distance depends on the following factors:

- vessel principal particulars and mooring/stationkeeping system type and performance of the vessel performing post-installation drilling;
- seabed well pattern;
- pullover direction relative to the seabed well pattern and to the direction of the environment;
- the various design pullover environments; and
- minimum vessel-to-vessel clearance requirements.

In particular, DnV POSMOOR has a 10-m clearance requirement, which was used for designing the mooring system of the MODU, Ocean Lexington, but a greater minimum clearance is often required by the operator's specification.

The design pullover environments include the following:

- maximum design environments for pullover winching operation;
- maximum design environment for pullover drilling operation;
- maximum design environment for pullover standby (typically, production is still maintained, but MODU is on standby); and
- maximum design pullover environment for the spar being in the pulled over position.

Pullover drilling is normally conducted off hurricane season with the spar workover rig demobilize to save rental of the rig as well as to reduce windage area. Typically only intact spar mooring systems in the pulled over position are considered for designing the spar mooring system.

Adequate winching capacity must be provided for pullover drilling. Pullover drilling requirements often dictate the use of an active mooring system and govern the following of spar mooring system designs:

1. minimum platform chain lengths inboard and outboard of the fairlead, and
2. tensioner (chain jack) capacity.

In addition, the adequacy of the spar mooring system design must be checked for the spar in its pulled over position subject to the maximum design pullover environment (for example, 100-year winter storm).

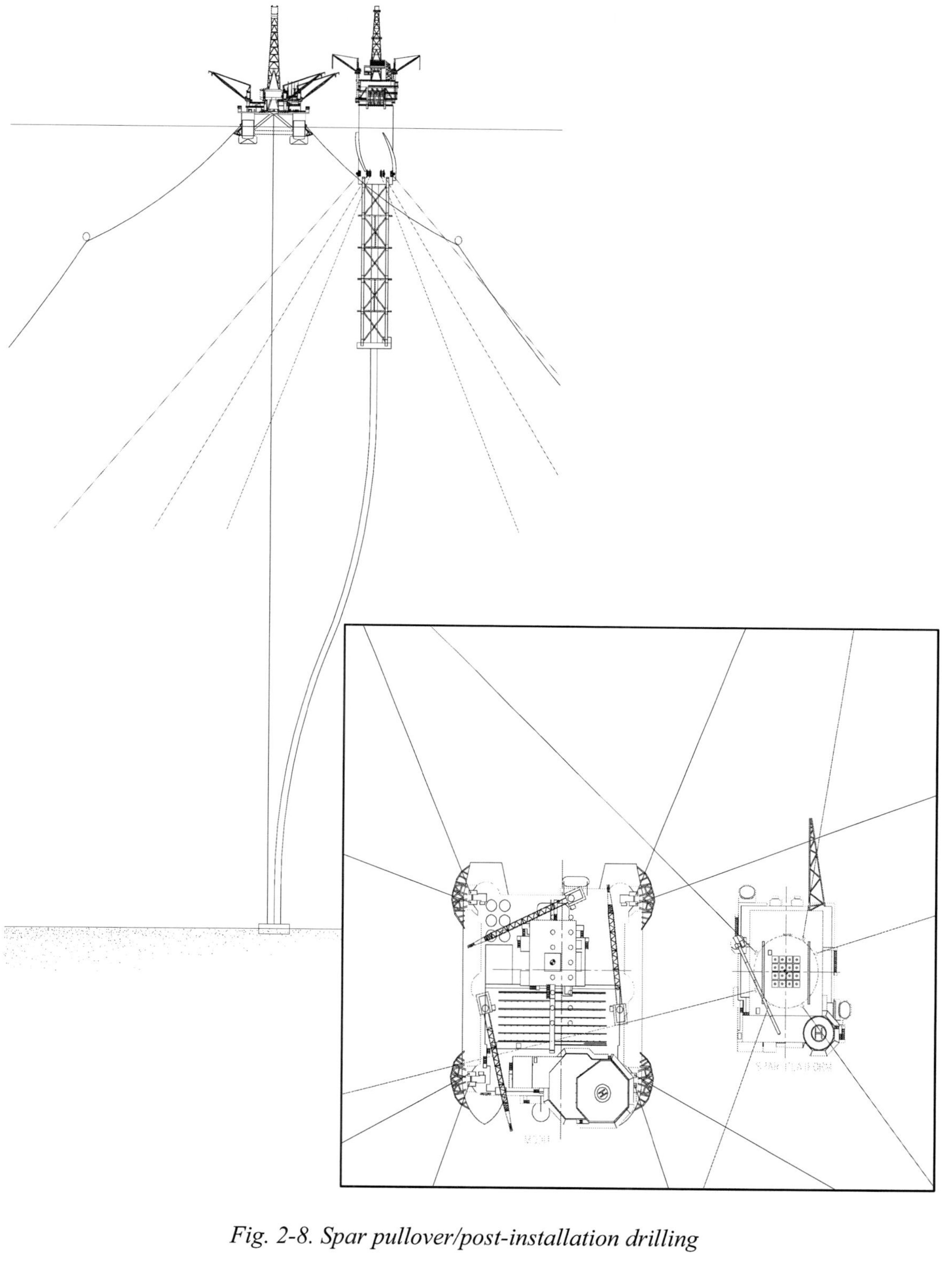

Fig. 2-8. Spar pullover/post-installation drilling

Catenary Riser Imbalance Loads

Apart from top-tensioned risers, a spar may be designed to support various catenary risers for injection, production from subsea wells, as well as for oil and/or gas export. Support of umbilicals for control of subsea wells also may be needed.

Depending on the number of risers, the hang-off arrangement from the spar hull, and the phasing scenario of the hang-off of the initial and any future risers and umbilicals a static imbalance load may be created, representing the resultant hang-off loads of all the risers for any given phasing scenario. While the spar hull's buoyancy can bear the vertical component of the imbalance load, its mooring system must provide resistance for the horizontal component.

This horizontal imbalance load affects the mooring system design. For example, differential line pretensions are needed to create an initial mooring system tension whose horizontal component exactly counter balances that of the static riser imbalance loads, leading to a skewed system configuration and/or line size. Experience indicates that this riser imbalance can be significant relative to the total environmental loads on the spar and can affect the system configuration and line size.

In-Place Mooring Line Change-out

Mooring system designs on existing spars have not yet considered in-place change-out as a design requirement. Although sparing of the mooring system is required to cover possible damage during installation (a full segment of jacketed spiral-strand wire rope is typically specified as a spare due to the long lead time in acquiring wire rope), no mechanism has been provided for in-place change-out of any mooring components when a line is damaged.

Incidences during installation of the Nansen and Bomvang Truss spar mooring system seem to have prompted changes in spar mooring system design. Subsea connectors will be used for some of the spars currently in the detail engineering and fabrication stage, such as the Gunnison truss spar mooring system design. For the Mad Dog and Red Hawk spar mooring system designs, use of subsea connectors can significantly facilitate the installation of the polyester mooring segments as well as the in-place retrieval of the test insert segments or the change-out of a line in case of damage.

Contract Model

The contract model affecting a spar mooring system design mainly relates to the contract for the mooring system installation. All existing spar mooring systems have been installed using an offshore crane vessel. This method of installation is largely determined by the contract model for the offshore installation work, which necessitates use of a crane vessel to lift of the deck offshore.

The installation method and capabilities of the installation vessel as dictated by the installation contract significantly affect the mooring system design. For example, the capability of offshore crane vessels, such as JRMD's DB-50 and HMC's Balder, is significantly greater than typical anchor handling vessels, such as the Cal Dive Aker Dove. As a result, large size lines and heavier anchor designs can be allowed.

In addition, the availability of the crane vessel may also affect the mooring system design as discussed in the next section.

Project Schedule

The project schedule can affect the spar mooring system design in a number of ways. For fast-track projects using concurrent design and fabrication, some conservatism in line size may have to be built into the procurement specifications if the mooring system design cannot be finalized before the order is placed. The procurement schedule may also affect vendor selection, resulting in different manufacturing constraints. Finally, the installation vessel schedule is constrained by the availability window for mooring pre-installation, which affects the mooring procurement schedule as well as the level of soil consolidation for calculating the holding capacity of the pile anchors. The holding capacity, in turn, affects the size of the anchor required.

To illustrate the point, a sample soil consolidation curve for a Gulf of Mexico field is shown in Figure 2-9. Soil consolidation and foundation setup must be completed about 180 days before installation to allow the soil to compress to two-thirds of its original undisturbed strength. Often, the level of consolidation drives the anchor size. In some cases, a somewhat less extreme environment can be justified to calculate foundation safety factors for the initial pre-commissioning phase prior to startup than for the in-place condition. For example, a lesser design environment (such as the 100-year winter storm) may be appropriate if the mooring hook-up and offshore installation/commissioning can be completed outside the hurricane season.

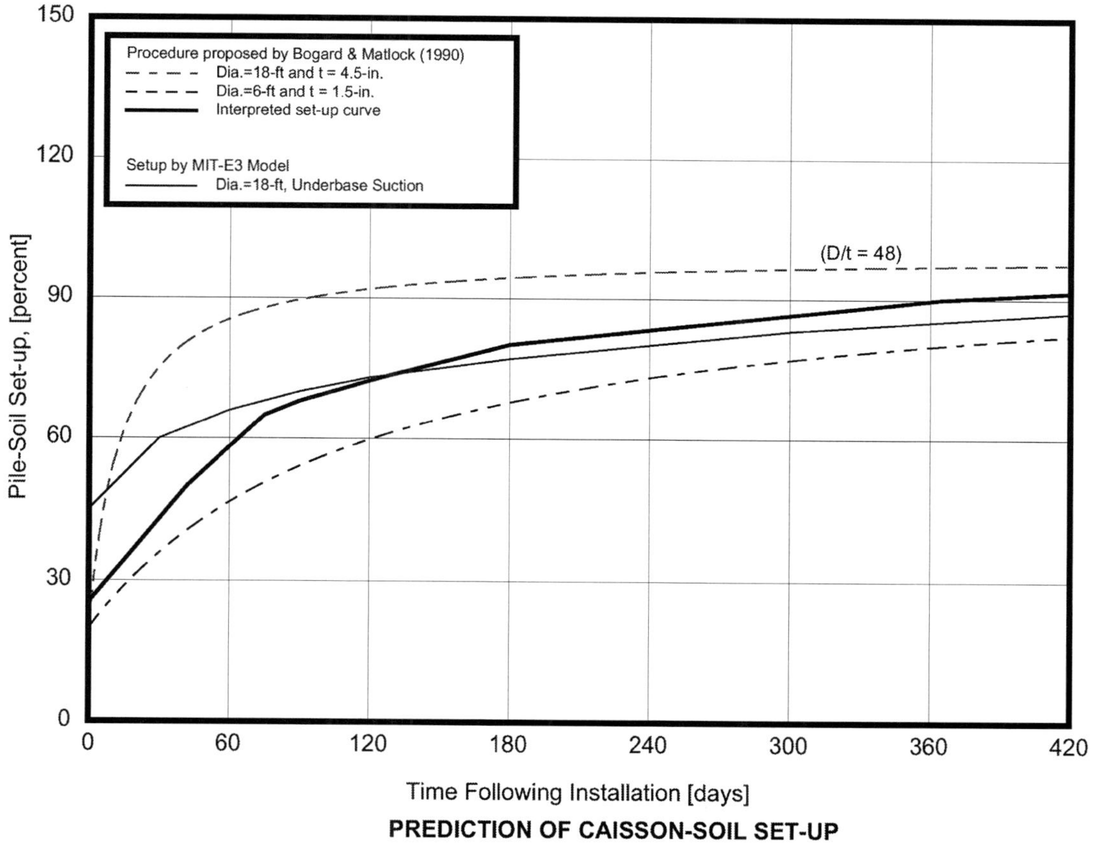

Fig. 2-9. A sample pile setup and soil consolidation curve

Design Constraints

The designer needs to be aware of the various constraints affecting the design. Some important ones are discussed in the following sections.

Manufacturing

Major manufacturing constraints to be considered in design and procurement of mooring system components include:

- type and grades of mooring lines in particular;
- tonnage capacity and delivery lead time;
- maximum line size capacity; and
- maximum wire rope spool length for transportation purposes.

It is important to verify the manufacturing limitations with potential qualified vendors to ensure feasibility of the design as well as better pricing. For example, the maximum chain size on order is 6.75 in. for the Holstein truss spar (a similar size chain is also being considered for the BP Atlantis semisubmersible platform). Bridon has a maximum transport load of 110 tonnes per spool, which corresponds to a segment length of about 3100 ft. for a 6-in. jacketed spiral strand wire rope. For reference, the maximum lengths for other sizes are shown in Table 2-2.

Table 2-2. Maximum Chain Lengths

Diameter		MBL	Unsheathed Strand weight	Absolute Max. Length at 110 tonnes		Allowances			Maximum specified length	
ins	mm	kN	Unsheathed kg/m	metres	feet	Sheathing set up	Test Piece	Tolerance 0.5%	metres	feet
4	102.5	10266	51.6	2132	6994	10	6	10.5	2105	6907
4 1/8	105.5	10867	54.7	2011	6598	10	6	9.9	1985	6513
4 1/4	108.0	11427	57.5	1913	6276	10	6	9.4	1888	6193
4 3/8	111.5	12129	61.0	1803	5916	10	6	8.9	1778	5835
4 1/2	114.0	12775	64.2	1713	5621	10	6	8.4	1689	5541
4 5/8	118.0	13594	68.4	1608	5276	10	6	7.9	1584	5198
4 3/4	121.5	14362	72.2	1524	4998	10	6	7.5	1500	4921
4 7/8	124.0	15073	75.9	1449	4755	10	6	7.1	1426	4679
5	127.0	15722	79.1	1391	4562	10	6	6.8	1368	4487
5 1/8	130.0	16480	83.3	1321	4332	10	6	6.5	1298	4259
5 1/4	133.0	17171	86.8	1267	4158	10	6	6.2	1245	4085
5 3/8	137.5	18272	92.5	1189	3901	10	6	5.8	1167	3830
5 1/2	141.0	19180	97.5	1128	3701	10	6	5.5	1107	3631
5 5/8	144.0	19867	101.3	1086	3563	10	6	5.3	1065	3493
5 3/4	146.5	20469	105.1	1047	3434	10	6	5.1	1025	3364
5 7/8	149.5	21385	109.3	1006	3302	10	6	4.9	985	3233
6	153.0	22070	114.4	962	3155	10	6	4.7	941	3087
6 1/8	156.0	22955	119.0	924	3033	10	6	4.5	904	2965
6 1/4	159.0	23835	123.6	890	2920	10	6	4.3	870	2853

System Installation

Key constraints include the maximum line weight and anchor weight/size the installation vessel can handle during installation.

For driven piles, existing hammer capability to install driven piles in ultra-deep waters beyond 5000 to 6000 ft. should also be confirmed with the hammer supplier (for example, Menck).

System Configuration

Various system configuration constraints may arise, depending on the project, including the following:

- Grounding the wire rope in the least loaded line is generally not allowed in intact mooring systems.
- Fairlead location can be conveniently accommodated in the hull structural design. Typically, the location is close to the bottom of the hard tank and in the same radial direction as the chain locker.
- Cut-off of VIV strakes for hull dry transportation must be carefully considered in terms of its azimuth location relative to float-off requirements, centerwell orientation, fairlead location, directions of the prevailing loop current (for GoM designs), and such.
- Cut-out of strakes in the way of the platform chain from the fairlead to the chain jack should be minimized.
- For practical reasons, the same-sized platform chain and ground chain are normally used.
- Mooring system layout should maximize the space for—and avoid interference with—the routing of catenary risers supported by the hull.
- Mooring system design and operation should be kept as simple and practical as possible. For example, variations in pretensions between lines in the same group and across groups should be kept small, where possible.
- Boat landing(s) should be downstream of the normal operating environments.

In loop-current driven designs, the spar surge/sway natural periods may need to be carefully tuned to minimize the lock-in VIV responses for most currents.

Mooring System Design

Overview of Design Requirements

Mooring system design for a spar is driven by

1. the global performance requirements of the spar, principally the maximum watch circles as well as maximum total of the resultant (pitch and roll) angular offset (sometimes referred to as maximum pitch) for the various design load cases (see "Load Case Matrix");
2. the requirement that the system design satisfies the required strength, fatigue holding capacity safety factors; and
3. the requirement that the system design satisfy the functional and operating specifications of the system as well as manufacturing, installation, and other constraints. Some of those constraints are discussed earlier (see "Design Drivers and Principal Considerations").

Design Safety Factors

Some of the past spar projects included a design survival environment corresponding to the hurricane environment with a 1000-year return period. Typically, an intact mooring system and hull condition is assumed, and the design safety factors for line tension and anchor holding capacity are allowed to reach 1.0. In this case, a set of load cases relevant to the survival environment also must be included in the load case matrix.

Load Case Matrix

An important part of establishing the mooring system design premise is to define the load case matrix. The load case matrix is an ensemble of analysis cases representing a permutation of the following design conditions:

- spar load conditions representing different system mass conditions at a given draft;
- spar draft conditions (for example, Kerr-McGee Neptune, Nansen and Bomvang spars are defined to be operated at a reduced draft when pulled over for post-installation drilling);
- spar rig mob/demobbed condition (if applicable), representing both system mass property and windage area changes;
- different spar positions (for example, drilling/completing different wells or pullover drilling using a MODU);
- different drilling and/or workover rig positions;
- different system damage scenarios (hull/mooring damage, riser casing leak, well shut-in, and such;
- different design environment conditions; and
- different riser staging scenarios representing how and how many initial and future risers (TTRs, SCRs, and other types of risers and control umbilicals) are designed to operate at different stages during the field design life.

Other design parameters affecting the mooring system design may also change, leading to additional load cases to include.

It is important to develop a comprehensive load case matrix that covers all the necessary load cases to be analyzed. It is equally important to group the load cases into three categories for the convenience of mooring system design:

1. Critical load cases—those cases that are considered to potentially govern one or more aspects of the mooring system design, global performance (offsets, pitch, deck accelerations, and such) riser, or other spar system design.
2. Non-critical load cases—those that are considered unlikely to govern any aspects of the mooring or other spar system design.
3. Non-transparent load cases—all the remaining cases, whose criticality cannot be readily discernable without performing the load case analysis.

Categorizing the load cases enables the analysis priorities to be set for different stages of the mooring system design (see "Design Procedure" for more discussion). The critical load cases will be the focus for the early stages of the mooring system design. Specifically, those cases in this category considered to be governing (for example, the 100-year design extreme environment and the 1000-year survival environment) need to be covered in the mooring sizing stage to define the mooring system design. The remaining critical cases can be covered in the preliminary mooring design to confirm the system design.

The mooring system design is not considered to be confirmed in principal without completing the analysis for the non-transparent load cases. Often some of the load cases may be equally critical, if not governing, based on the analysis results. These should be regrouped into the critical load cases. Additionally, some of the load cases originally deemed critical may be down-graded. By the end of this mooring design cycle, the critical and non-critical load cases are regrouped and more clearly defined, and the number of non-transparent load cases is reduced to a small group. This progression of mooring analysis sharpens and maintains the design focus.

The mooring system design may now need some fine-tuning based on the analyses of the non-transparent load cases. By the end of these analyses, the design should be verified in principal. However, prior to the fine-tuning effort, the engineer may have had to select the line size (line diameter) for mooring procurement, which is an important engineering milestone.

The non-critical cases are usually just complementary analysis, which are completed and checked in the detailed system design as a matter of procedure. Mooring system specifications for procurement are typically developed as the analysis for these load cases are being completed.

Design Procedure

The spar mooring system design involves the following principal steps done at different stages of the project:

- review design basis;
- formulate design premise;
- size the mooring system;
- conduct preliminary analysis;
- perform detail analysis and component specification and procurement; and
- conduct post-installation analysis.

Review of Design Basis

Conceptual field and mooring system layouts need to be developed together with identifying the following:

1. bathymetry and existing subsea installation geohazards potentially affecting the mooring system layout and design;
2. environment directionality and platform orientation;
3. fendering and supply boat mooring arrangement needs;
4. direction(s) of design extreme environment affecting mooring system load-sharing or hull VIV suppression considerations;
5. catenary risers and subsea flowline/pipeline layout requirements;
6. wellbay pattern at the seabed level;
7. pullover direction of the spar for post-installation drilling as well as the MOPDU mooring footprint requirements (if relevant);
8. feasible options for different mooring system configurations; and
9. soil conditions and anchor design options suitable for the soil conditions.

In particular, a quick mooring system sizing can be performed using a preferred (or base case) mooring system configuration to estimate the mooring line size, scope, and horizontal spread (rough anchor footprint); maximum tensions; and anchor holding capacity. In most cases a rough estimate of the mooring system design can be developed based on past design experience. This can then be used to update the field and mooring system layout.

Under normal circumstances, a preliminary system optimization may be performed to obtain preliminary mooring system sizing and costing for a number of feasible system configurations subject to various concept-level design constraints. The system variables include line grouping pattern, number of lines, line size, scope and line composition, and pretension. The objective is to minimize the total system installed cost.

Typically, certain simplifications and assumptions are made in the mooring sizing at this stage. For example, the water depth and line composition for all mooring lines may be used. The seabed bathymetry is ignored, and an omni-directional (and collinear) environment may be used. Additionally, the designer may have to make some assumptions about aspects of the project execution model affecting the mooring system procurement and installation.

Normally, the decision to use (more or less) an equal spread mooring pattern versus a grouped mooring pattern depends on the efficiency of system load sharing (in intact and damaged conditions), directionality of the environment, and clearance/interference considerations, among other things. A general conclusion is difficult to draw.

Trade-off of mooring line size versus the number of mooring lines is one component of the system optimization study. Intuitively, the line size will increase when fewer lines are used. Past project experience shows that using fewer mooring lines often reduces total system installed cost.

Line composition versus scope (total line length) is usually a secondary system optimization in which longer wire length is traded with shorter chain length (typically ground chain). The line size is usually held constant. While the total scope may change, the cost variation is usually of secondary impact, and the optimization may be omitted at this stage of the mooring design. For taut-leg mooring systems, the scope ratio (the total line length divided by the water depth) is mostly driven by the maximum offset limit and water depth.

Variation in pretension also affects the line size, composition, and scope. The pretension typically varies from 15 to 20 percent of the mooring line breaking strength for the design extreme environment condition. To make the most use of the mooring lines, a high pretension is preferable as long as the mooring line fatigue (particularly the chain fatigue) and chain wear life is acceptable. For normal drilling operations, which are considered to be temporary, higher line pretension may be used depending on the offset limit and chain jack capacity. Nonetheless, it is important to account for the impact of higher line pretensions under these conditions for fatigue-critical mooring system design.

Another set of mooring system optimizations is steel versus polyester mooring systems. The Horn Mountain and Mad Dog truss spar designs have explicitly considered both types of mooring designs. The cost-effectiveness of the system, however, is usually not the only consideration.

Design Premise Definition

A comprehensive design premise document must usually be developed in complement with spar global performance. As a document subject to classification review, it must at least cover the following aspects of the mooring system design:

- general system design requirements;
- project-specific design requirements;
- environment and geo-technical information;
- define design input information;
- design life;
- codes and standards;
- design criteria, including factors of safety, and system performance criteria;
- key constraints and assumptions;
- load case matrix; and
- software.

Mooring System Sizing

This is an iterative process, in which a preliminary mooring system configuration, mooring line size, and composition are developed iteratively, usually based on simplified analysis and/or for a set of governing load cases. This preliminary mooring system design satisfies the design requirements. The closeness of the preliminary design to the final design depends on a number of factors, including the designer's experience in identifying the governing load cases to be covered.

Simplified analysis is often performed at this stage to develop a preliminary design quickly. In addition to simplifying the method used, software and modeling sophistication and the number of load cases can also be reduced.

Typically, nominal line lengths with catalog line properties are used. Connectors are ignored, and fairlead positions and line departures are often idealized.

The anchor is assumed to be located and fixed at the nominal burial point (sometimes also referred to as the mudline point) so that the inverse catenary of the buried portion of the ground chain is neglected (except the elasticity as represented by a correction to the axial stiffness (EA) of the ground chain above the mudline). The same can be said about the platform chain, which is assumed to be fixed at the fairlead. As such, an adjustment is made to the EA of the platform chain outboard of the fairlead to account for the elasticity of the platform chain from fairlead to stopper. Similarly, to arrive at the line top tension at the stopper (nominally at the chain jack), the line top tension at the fairlead as obtained directly from the analysis is corrected to account for weight of the platform chain from fairlead to stopper.

Preliminary Design

This phase of the mooring design covers the remaining critical load cases as well as those non-transparent load cases. The analyses in this phase seek to verify and fine-tune the preliminary design to fix the line size as early as possible so that the on-vessel component design can begin and firm input data can be generated for the anchor design.

Based on the preliminary spar deck layout, actual on-vessel mooring component locations are factored into the analysis, which must still rely on catalog mooring line properties. At the same time, the categorization of the load cases must be updated based on the analysis results.

At this stage, mooring installation requirements are better defined and can be progressively covered in the analysis verification.

Detail Design and Component Specification and Procurement

At the detailed design stage, the strategy and procedure may differ somewhat depending on the project-specific circumstances. However, a main focus is revisiting the governing cases using more advanced/accurate analysis software to ensure important features affecting the design are adequately captured in the earlier analysis, while at the same time completing the non-critical case runs. Once again, a two-tier approach to detail design is often necessary in the early stage of project transition from preliminary design to detail design. For example, a time-domain analysis is often performed for the governing load case or for cases identified while completing the detail analysis for the non-critical load cases.

Another focus of the detail engineering is to support and finalize the system and component specifications to request proposals from vendors. Requests for quotes (RFQ) may be initiated based on preliminary mooring system design specifications developed in the later stage of the preliminary design. Specifications are updated and purchase orders placed at this stage. For components with long-lead delivery time (such as chain jacks and jacketed spiral strand wire), procurement may be executed by setting the lines size and/or tension budget for the mooring system design.

Post-Installation Analysis

A post-installation analysis is done upon receipt of the actual pre-installed mooring system data. The purpose is (1) to make sure that the as-installed mooring system is acceptable and (2) to determine the so-called zero link, the chain link to be held up by the stopper when the spar is in the design target zero-equilibrium position under the design pretension.

The actual anchor locations and mooring line as-installed lengths, which account for the lengths of individual segments and connectors, will be used to determine the zero-link location for each line as well as the as-installed spar zero-equilibrium position. Because each stroke of the tensioner (assuming linear tensioners) counts for two chain links, a platform chain length equal to the discrete number of links must be used in the calculation. As a result, the actual mooring line pretensions will differ (albeit slightly) from the target pretensions, and the spar actual zero-equilibrium position may be offset slightly from its target position.

Method of Analysis

The method of analysis and the software used depends on the stage of the mooring system design, among other things. At each stage of the analysis, the purpose of the mooring system design, the accuracy level, and the analysis efficiency required differ. The methodology and the software used also vary in the degree of sophistication. These can be grouped according to one or a combination of the following:

- quasi-static versus dynamics;
- frequency-domain versus time-domain; and/or
- coupled (hull/mooring/riser) versus uncoupled.

General descriptions of various mooring analysis methods abound in the open literature. Issues and necessary background associated with use of any specific software are usually covered in the software's user manual.

Design Safety Factors

Design safety factors include factors for strength and fatigue of the mooring lines, connectors, and anchor holding capacity.

Mooring System

Mooring line safety factors are specified in API RP 2SK.

Connectors, on-vessel equipment, and foundations must be designed to normal structural design tolerances for allowable stresses and fatigue life.

Component	Strength	Fatigue
Mooring Line	1.67 intact/1.25-damaged	10 × design service life
Connectors	1.25 intact/1.0-damaged	10 × design service life
On-Vessel Components	110% MBL	3 or 10 × design service life depending on if fully inspectable, and replaceable
Foundations	1.25 intact/1.0-damaged	3 or 10 × design service life depending on if fully inspectable, and replaceable

Anchor Holding Capacity

The anchor holding capacity is determined by the load and resistance factor design of API RP 2A-LRFD.

The foundation meets the criterion if

$$\gamma.F \; [\; \Sigma \; \phi_i.R_i \qquad (2\text{-}1)$$

$$\gamma.F \; [\; \phi_s.R_s + \phi_w.R_w$$

where

γ = load factor
F = nominal load
ϕ_i = component resistance factor
R_i = component resistance
ϕ_s = resistance factor associated with soil resistance
R_s = soil resistance including skin friction and reverse end bearing
ϕ_w = resistance factor associated with pile weight (0.95)
R_w = pile weight in seawater less soil buoyancy due to penetration

	Load Factor	FOS_{equiv}
Temporary Conditions (Intact):		
Load Factor =1.30		
Soil Resistance Factor ($\theta < 35^{\sim}$) =	0.90	1.44
Soil Resistance Factor ($\theta / 35^{\sim}$) =	0.87	1.50
Permanent In-place Condition (Intact):		
Load Factor =1.30		
Soil Resistance Factor ($\theta < 35^{\sim}$) =	0.80	1.63
Soil Resistance Factor ($\theta / 35^{\sim}$) =	0.65	2.00
Permanent In-place Condition (1-line damaged):		
Soil Resistance Factor ($\theta < 35^{\sim}$) =	1.00	1.30
Soil Resistance Factor ($\theta / 35^{\sim}$) =	0.87	1.50
Other Geotechnical Factors of Safety		
Embedment pressure	1.5	
Extraction pressure		1.5

The following safety factors are shown as an example of driven pile designs used with ±5° verticality tolerance and ±7° orientation tolerance:

Environment Condition	Mooring Condition	Lateral Soil	Axial Soil	Pile Steel
Limit Operating	Intact	2.0	2.5	0.75Fy
100-Year Return	Intact	1.6	2.0	1.33 × 0.75Fy
100-Year Return	Damaged	1.2	1.5	1.33 × 0.75Fy

Hardware Selection

Chain

Studlink Versus Studless

The chain used for offshore applications includes studlink and studless chains. The industry has had extensive experience with the use of studlink chains associated with MODU mooring systems, where studlink chains are often preferred due to ease of handling.

However, studlink chains are heavier than studless chains at equivalent breaking strength (though slightly stronger for the same line size). In addition, past industry experience has indicated a significant level of stud loosening due to high stress concentration and fatigue. For permanent mooring system design with a taut-leg configuration, studless chain is generally selected. Discussions below are mainly focused on studless chains.

In deep water, the self-weight of the mooring chain mooring system produces a steep catenary profile and a softer line with a high vertical load component. The vertical component of the mooring loads results in a reduction in spar payload capability. Additionally, deepwater mooring lines with a large self-weight of can create deployment problems.

Chain Grades

Chains are graded based on their breaking strength as ORQ, ORQ+20%, RQ3 (R4), and RQ4 (R4). R4 studless chains are of primary interest for deepwater spar mooring system design. Small-diameter tail chains inside chain lockers are often of lesser grade studlink for ease of handling.

Per DnV specification, proof load and breaking loads for a R4 studless chain are given in terms of chain diameter (d) as:

Proof load: $0.0192 \times d^2 \times (44—0.08d)$ (kN) (d=nominal diameter in mm)
Breaking load: $0.0274 \times d^2 \times (44—0.08d)$ (kN)

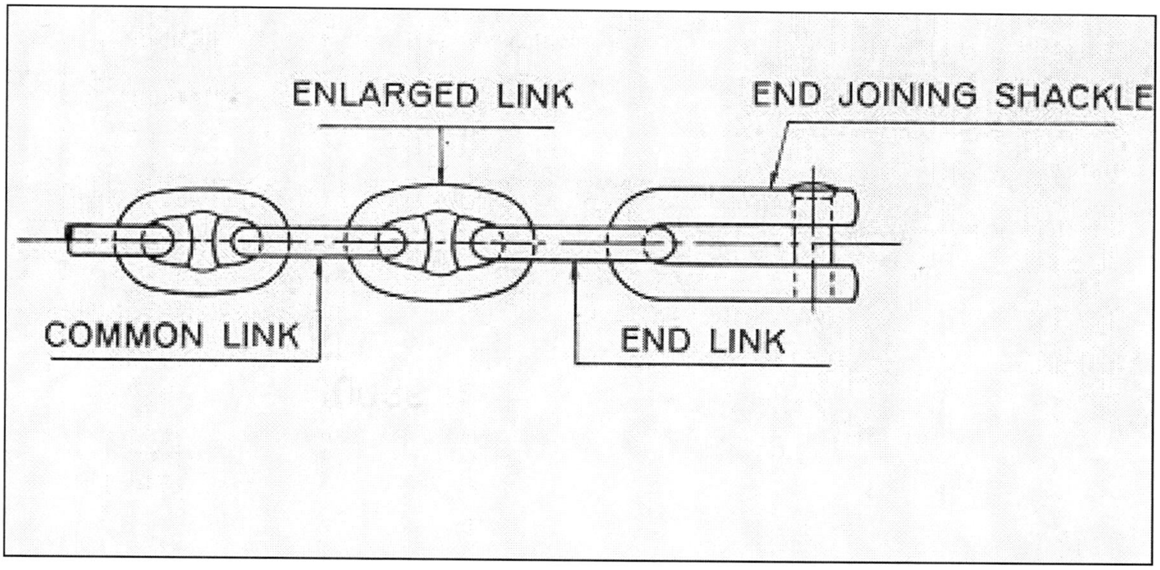

Fig. 2-10. The types of chain links

Chain Links

There are three types of chain links—common link, enlarged link, and end link, as shown in Figure 2-10.

The bar stock of a common link has the same diameter as the nominal chain diameter, whereas the bar stock diameter is slightly greater for the enlarged link (10 percent) and end link (20 percent).

Chain Specification

Key technical aspects of chain specification cover the following:

- material and mechanical properties;
- strength and fatigue life;
- manufacturing and testing;
- inspection;
- dimensional tolerances; and
- classification and certification.

Wire Rope

Wire ropes have a much higher strength/weight ratio when compared with chain. Wire moorings produce a significantly shallower catenary profile, which results in higher horizontal restoring forces.

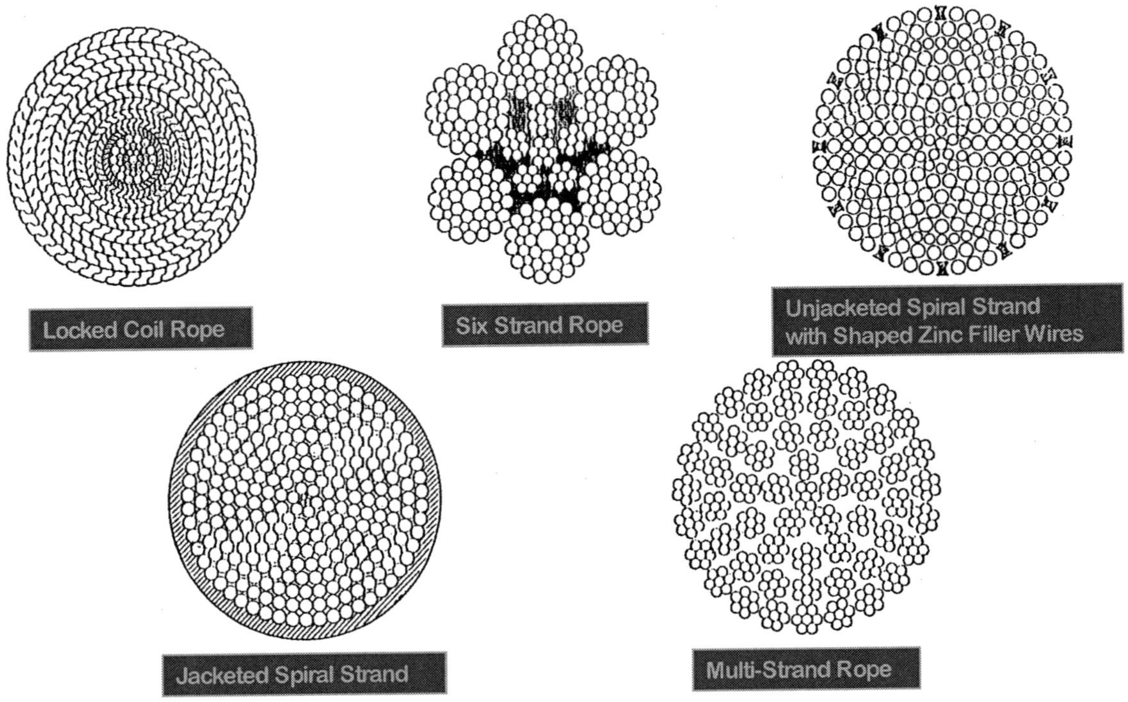

Fig. 2-11. Different types of wire ropes used in the offshore industry

Life Expectancy of Wire Rope Construction Types	
Wire Rope	**Life Expectancy**
Galvanized 6-Strand	6 to 8 Years
Galvanized Unjacketed Spiral Strand	10 to 12 Years
Galvanized Unjacketed Spiral Strand with Zinc Filler Wires	15 to 17 Years
Galvanized Jacketed Spiral Strand	20 to 25 Years
Galvanized Jacketed Spiral Strand with Zinc Filler Wires	30 to 35 Years

Figure 2-11 shows the types of steel wire rope construction commonly used in the offshore industry with their expected life expectancies. Jacketed spiral strand wire rope is of particular interest in spar mooring design and is covered here.

Typical line properties of jacketed spiral strand wire ropes can be easily obtained from manufacturers such as Bridon.

Fairlead

There are essentially two types of fairleads potentially applicable to spar mooring system—the bending shoe type and rotary type as shown in Figure 2-12.

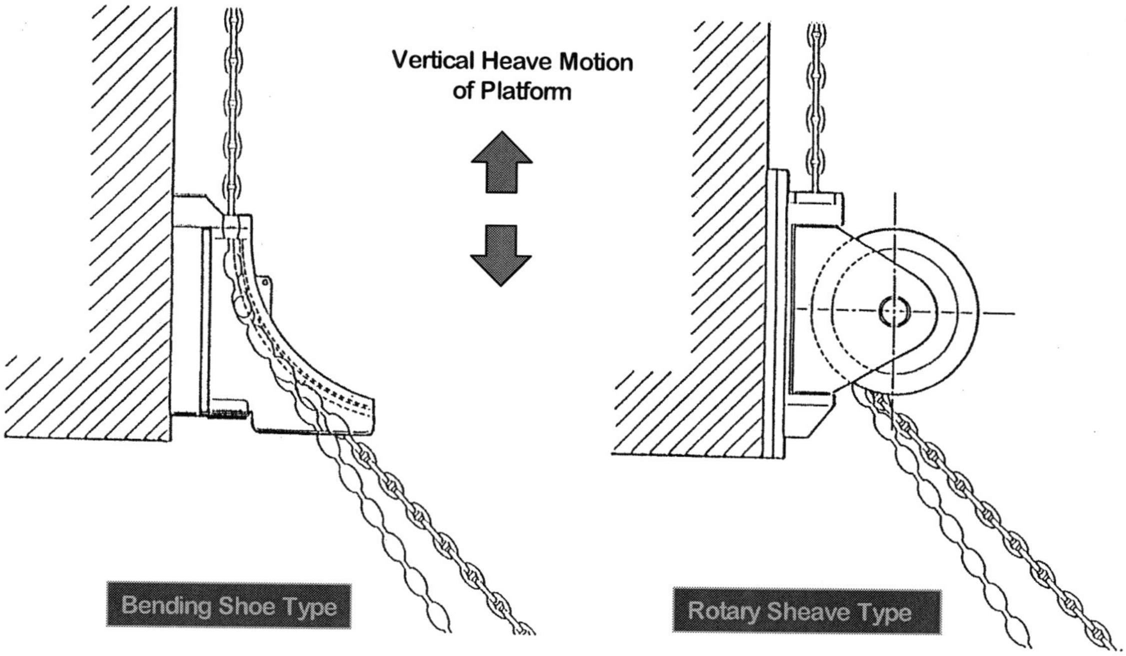

Fig. 2-12. Two types of commonly used fairleads

The bending shoe type fairleads are usually simpler and cheaper. However, due to premature wearing of the liners on earlier permanent mooring systems, rotary fairleads are usually used for permanent systems.

Based on field experience, rotating fairleads are preferable to bending shoe fairleads. Typically, five- or seven-pocket rotary fairleads are used to increase line-contact length. This reduces the long-term wearing and fatigue damage of the platform chain near the fairleads (Fig. 2-12), which are located near the bottom of the hard tank.

Rotary type fairleads usually have a large azimuthal sweep angle range of ±60°, which although good in accommodating yaw motions of the spar, entails cutting a large hole around the point on the strakes where the mooring lines run up from the fairlead to the stopper to avoid the lines chafing on the strakes. These cut-outs have been shown to significantly reduce the effectiveness of the strakes in suppressing the hull VIV motions. As such, it is important to minimize cut-outs, which normally decrease with the mooring line sweep area rising from the fairleads. Devices to cover the cut-out after the mooring lines are attached to the spar are being studied. Some fairlead manufacturers have been exploring designs to eliminate the sweep of the mooring lines above the fairlead and allow minimal strake cut-out. Mooring line hook-up procedures for these devices might require pre-rigging a message line through the holes.

Fairlead selection must also take into account whether an active or passive system is required. For an active system, frequent winching of the mooring system alleviates the need to adjust the chain links engaged in the fairlead wildcat sheaves. Although this is not the case with passive systems, it may still be necessary to provide a way to even up the mooring lines to avoid excessive wearing of the chain links engaging the fairleads. The mooring system design must accommodate the number of links that need to be rolled in or paid out. The interval and sequence for any preplanned adjustment need to be carefully worked out. Alternatively, larger fairleads may be used to increase the wrap-around angle and, therefore, the number of chain links engaged in the wildcat sheave of the fairleads to increase the contact area and reduce the long-term wearing and fatigue to an acceptable level.

The fairlead size is driven primarily by the mooring line size and the level of line tensions expected. Other important factors include requirements for in-service inspectability and replaceability, design and operational life, and fatigue and wear. For the latter, chain cyclic loads and travel loads/distances as well as pivoting loads and angle ranges will be needed.

Some ongoing spar projects require that the fairleads be replaceable in place. Pull-up of the fairleads through cut-outs on strakes has been considered an option for some earlier spar designs, but alternative procedures for in-place fairlead replacement would be necessary if the cut-outs were to be minimized for effective hull-VIV suppression.

Key technical specifications of fairleads typically include the following aspects:

- platform chain size to be accommodated and breaking strength (minimum breaking load [MBL]);
- fatigue and wear design life;
- nominal line pretension;
- nominal wrap angle and wrap angle range;
- pivoting (azimuthal sweep) angle range;
- 100 percent of the platform chain MBL for a range of wrap and pivoting angles for fairlead and foundation design;
- lifting and handling requirements;
- in-service inspectability and removability and specific parts to be designed for removal;
- materials and properties;
- means of lubrication for bearings and other thrust-bearing surfaces; and
- coating and corrosion protection requirements.

For some floating structures designed with a permanent but passive mooring system, the fairleads and on-vessel tensioning system have been eliminated, and each is replaced with a padeye. Line tensioning is done off-vessel during installation. This type of design can significantly reduce mooring system cost if it is acceptable from an operations, risk, and cost of future mooring line replacements perspective.

Tensioner

To date chain jacks have been used exclusively on all existing spar mooring designs due to their compact size and light weight compared with winches of the same capacity (the spar deck where the chain jacks are located is valuable real estate and is normally very congested).

Chain jacks are usually powered hydraulically. Depending on the number of mooring lines and level of power requirements, one or two hydraulic power units (HPUs) may be required to operate the chain jacks.

The tensioning capacity of a chain jack is usually driven by the operating requirement for active adjustment of the system line tensions and spar positions for on-vessel or pull-over drilling and well completion. This type of winching operation is normally conducted in a moderate environment.

Anchors

A variety of anchoring options can be found in offshore mooring applications, such as:

- drag embedment anchors;
- driven, jetted or drilled and grouted piles;
- suction embedment anchors;
- deadweight anchor; and
- vertically loaded anchors (VLAs).

Conventional high-holding-capacity (HHC) drag embedment anchors, such as the Vryhof or Bruce-type, drag-embedded anchors, have been used on permanent mooring systems. However, they are not suitable for taut-leg mooring system design on spars because of their limited resistance to vertical loads (max anchor uplift angle limited to 10°). They are also disadvantageous in that (1) they require proof loading, which can be difficult and/or costly because of the level of proof-tension required, and (2) an accurate positioning is difficult to achieve after proof tension due to the nature of drag embedment. Driven and suction piles are of primary interest and are covered here.

Suction Anchors

Suction anchors, like conventional piles, are individually designed and fabricated for specific soil conditions and mooring tensions. They are, therefore, relatively high cost items. They may be designed for use in semi-taut or taut leg mooring systems.

They also require careful alignment during deployment and are of limited use for hard soils, such as corals or compacted clays. Correct embedment of suction anchors cannot be ensured prior to deployment unless there has been a comprehensive and accurate assessment of soil conditions to the design embedment depth. As with driven piles, security of the anchors is good when correctly embedded. They can be designed to be recoverable in theory, but as they are designed to be location specific, there may be little value in recovery. Uninstalling and reusing suctions has yet to be done in practice.

Extreme water depth poses difficulties for the use of hydraulically driven evacuation pumps and electric or WROV-driven units may be required.

Figure 2-13 shows a typical suction anchor—an upended suction anchor being lifted off the installation vessel with the ground chain connected. Notice the padeye position on the suction anchor.

Driven Piles

Because of their small size relative to suction piles and lower installed cost, driven piles have been preferred in spar installations in water depths where existing driving equipment can be used.

Driven pile is considered to be proven technology for up to 4000 ft. and has been extensively used on fixed platform and TLP mooring foundations. According to Menck, its existing underwater hammers are capable of driving piles up to 6000 ft. ConocoPhillips' Magnolia TLP design, soon to be installed in 4800-ft. water depth in the Gulf of Mexico, uses driven pile foundations.

Because of the significant lateral loads that driven piles for anchoring a spread mooring system are designed to resist compared with those used as TLP mooring foundations, driven piles for spread mooring systems are stubbier (much lower length/OD ratio) than those used as an TLP mooring foundation. Figure 2-14 shows a typical driven pile. For comparison purposes, typical LOA (length overall)/OD (outside diameter) ratios are 5 for suction piles, 25 for driven piles for spread mooring systems, and 50 for TLP foundation piles.

Connectors

Connectors are used to join different sections of a mooring line and to connect the mooring line to the anchor. They include connectors between the following:

- chain to wire;
- wire to wire;
- ground chain to anchor pile; and
- subsea connectors.

Chain to Wire Connection

Figure 2-15 shows the two types of terminations typically used for jacketed spiral strand wire ropes. Figure 2-15 also shows connections between two wire segments with open or closed sockets. For wire ropes with an open-socket termination, the rope-to-chain connection can be accomplished through a triplate or a "dog born" with a lifting padeye as shown in Figure 2-16. For wire ropes with a closed-socket termination, it can be designed to be directly connected to the D-shackle on the end of the chain. Figure 2-17 shows a wire rope segment with a closed-socket termination being readied for connection.

Note the anode bars attached onto the socket and its hard collar for holding the socket in position while being connected during installation.

Fig. 2-13. A suction pile foundation for a spread mooring system as it is lifted

Fig. 2-14. A driven pile foundation for a spread mooring system

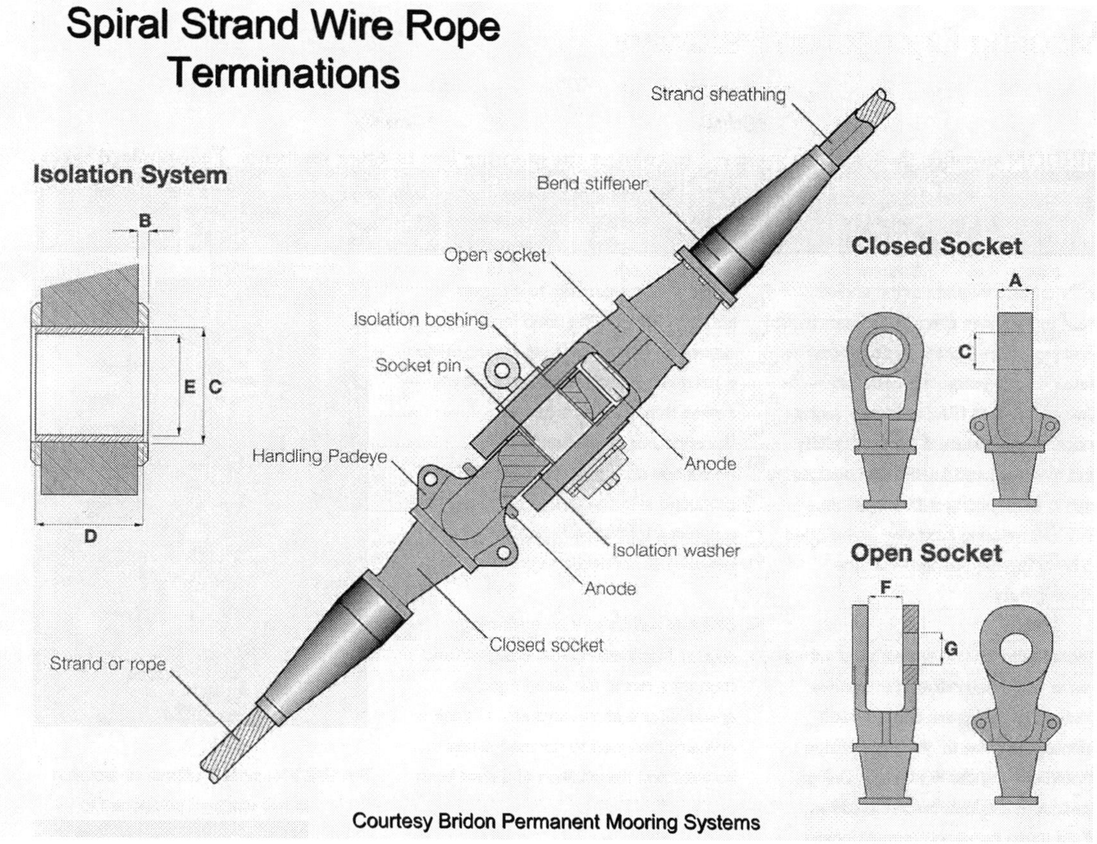

Fig. 2-15. Wire rope terminations

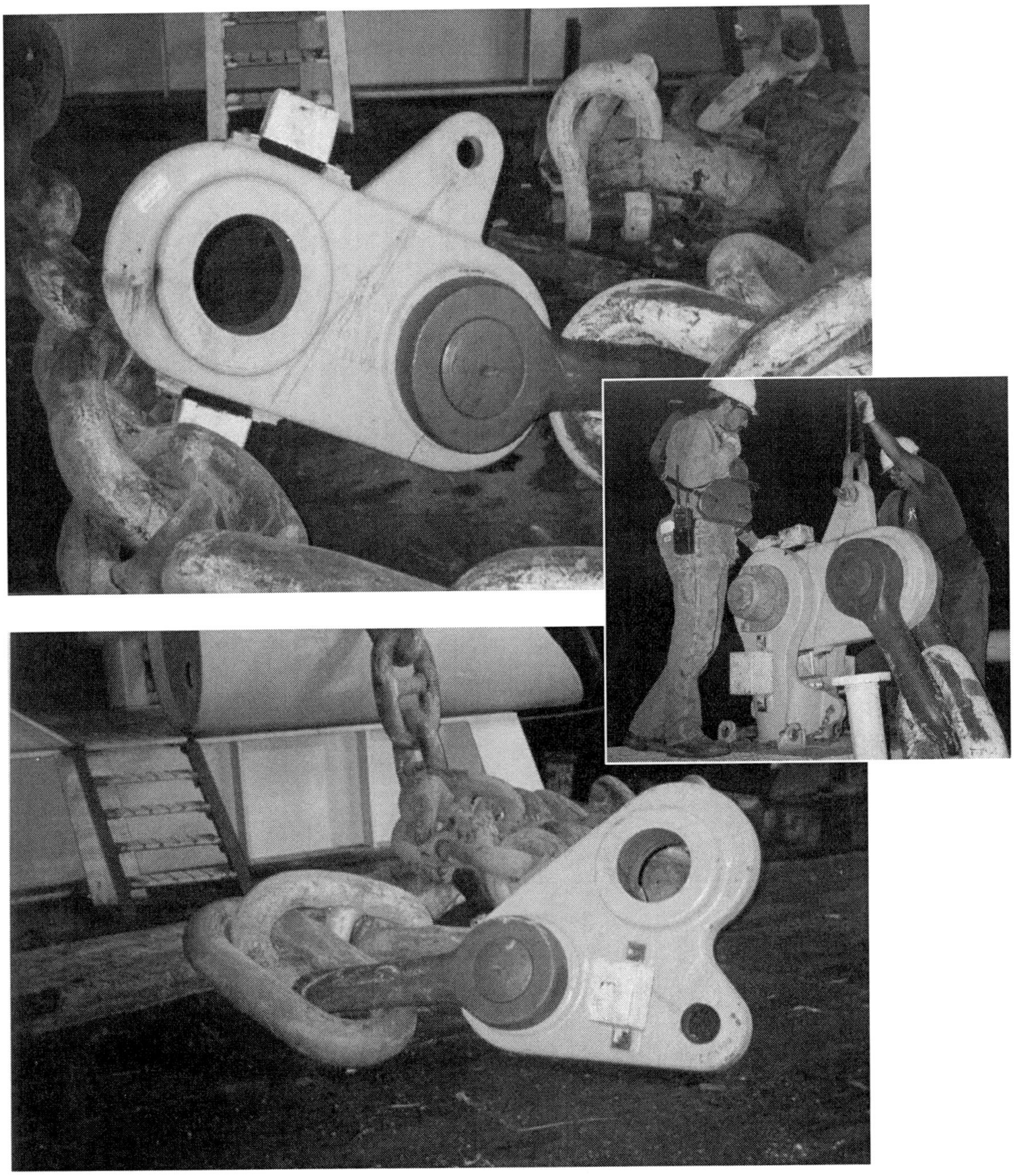

Fig. 2-16. Connection of chain to wire rope with an open-socket termination

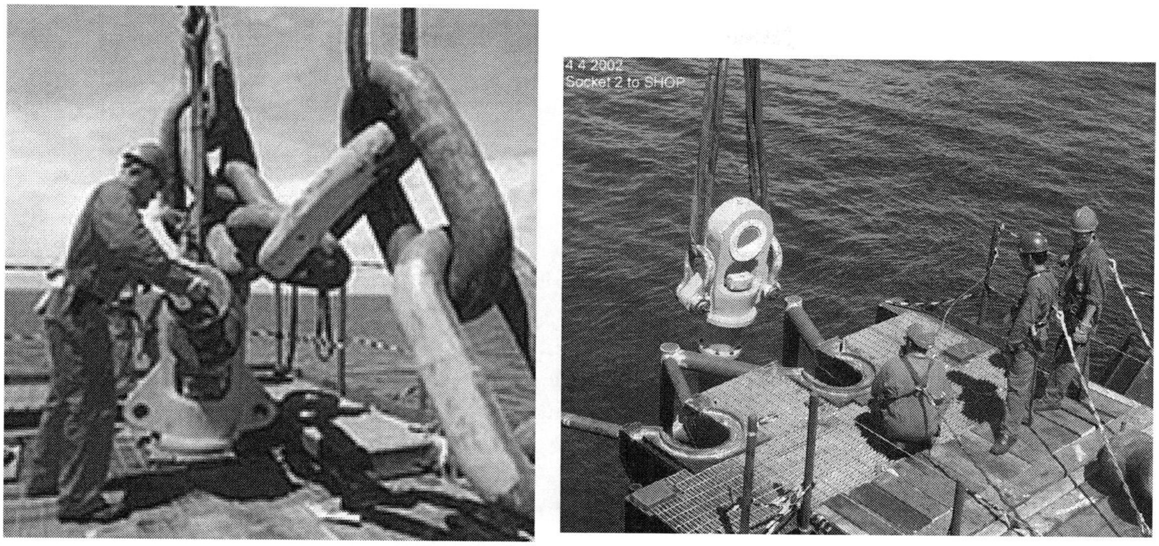

Fig. 2-17. A wire rope segment with a closed-socket termination ready to be connected

Wire-to-Wire Connection

Figure 2-15 shows a wire-to-wire connection in which a closed socket on one wire segment is fitted directly into the open socket of another segment. An alternate to two closed-socket (usually cheaper than an open socket) terminations on both wire rope segments is to use a connecting chain segment (see Figure 2-17) made of 2 D-shackles (one at each end) with several chain links (end links) in the middle.

Ground–Chain-to-Anchor Connection

Typically, an enlarged D shackle (anchor shackle) is used to connect the ground chain to the anchor padeye as shown in Figure 2-13.

Subsea Connectors

For ease of installation, line retrieval/replacement, or simultaneous installation of polyester wire ropes subsea connectors are often used.

The lower end (female part) of a subsea connector is connected to the upper end of the ground chain, and it is pre-installed with an anchor. The other end of the connector is at the lower end of the wire segment connected to the ground chain. The two parts of the subsea connector are then mated subsea via ROV assistance. This procedure depends on the connector type, among other things, and needs to be worked out specifically based on the mooring system design.

The two types of subsea mooring connectors used so far are the Delmar connector (used on MODU moorings) and BSW's ballgrab connector. Details can be obtained from the connector manufacturers.

Installation

General

Although several options are available for installing spar mooring systems, spar mooring installation has so far fundamentally used the methods and procedures dictated by the contract model for the whole spar installation, which is driven by the deck-lift operations offshore.

As shown in Figure 2-18, mooring installation typically starts with pre-installing the anchors (normally driven piles or suction piles) and prelaying the mooring lines. The ground chain section is usually connected to the anchor prior to lowering the anchor. Without using a subsea connector, either the lower section or all of the mooring line must be lowered with the anchor and secured using the line and a messenger wire during anchor setting. Once the anchor is driven to its design penetration, the whole line is laid on the seabed. It is recommended that installation start with the least-loaded line (the least critical line) in case of damage.

In very deep water, the mooring line and anchor can be very heavy and often become the limiting factor in mooring installation. Use of a subsea connector increases the installation flexibility. With subsea connectors, it is also possible to improve the anchor setting efficiency by lowering the piles in batches to their self-penetration depth and then batch driving the piles to their design penetration either by suction pressure or with an underwater hammer.

Typical anchor positioning accuracy is about a 10-ft. radius, depending on the type of positioning system used, the type and size of the anchors, and the capabilities of the installation vessel. Anchor verticality of 5° and orientation tolerances of 5° to 7° are considered to be realistic.

Installation records need to be maintained. As-installed information must be recorded and fed back to the design for verification of acceptability.

Driven pile installation is similar. Figure 2-14 shows a driven pile being lowered with the hammer and hydraulic umbilical.

Deployment of Spiral Strand Wire Ropes

Two key factors of jacketed spiral strand wire rope are the torque balance and compressive pressures on the sheathing during transportation, handling, and installation. Therefore, the messenger lines used need to be (near) torque balanced. Otherwise, a means of resisting messenger wire rotation must be incorporated into the installation procedure.

Damage to the sheathing can occur due to physical scrapping or over pressurizing of the jacketing during installation. Figure 2-19 shows an example of damage to jacketing and the wire itself.

Empirical formulae are available for estimating the maximum allowable pressure (for example, the *Offshore* magazine issues listed as references). However, it is highly recommended to consult the manufacturer for specific guidance.

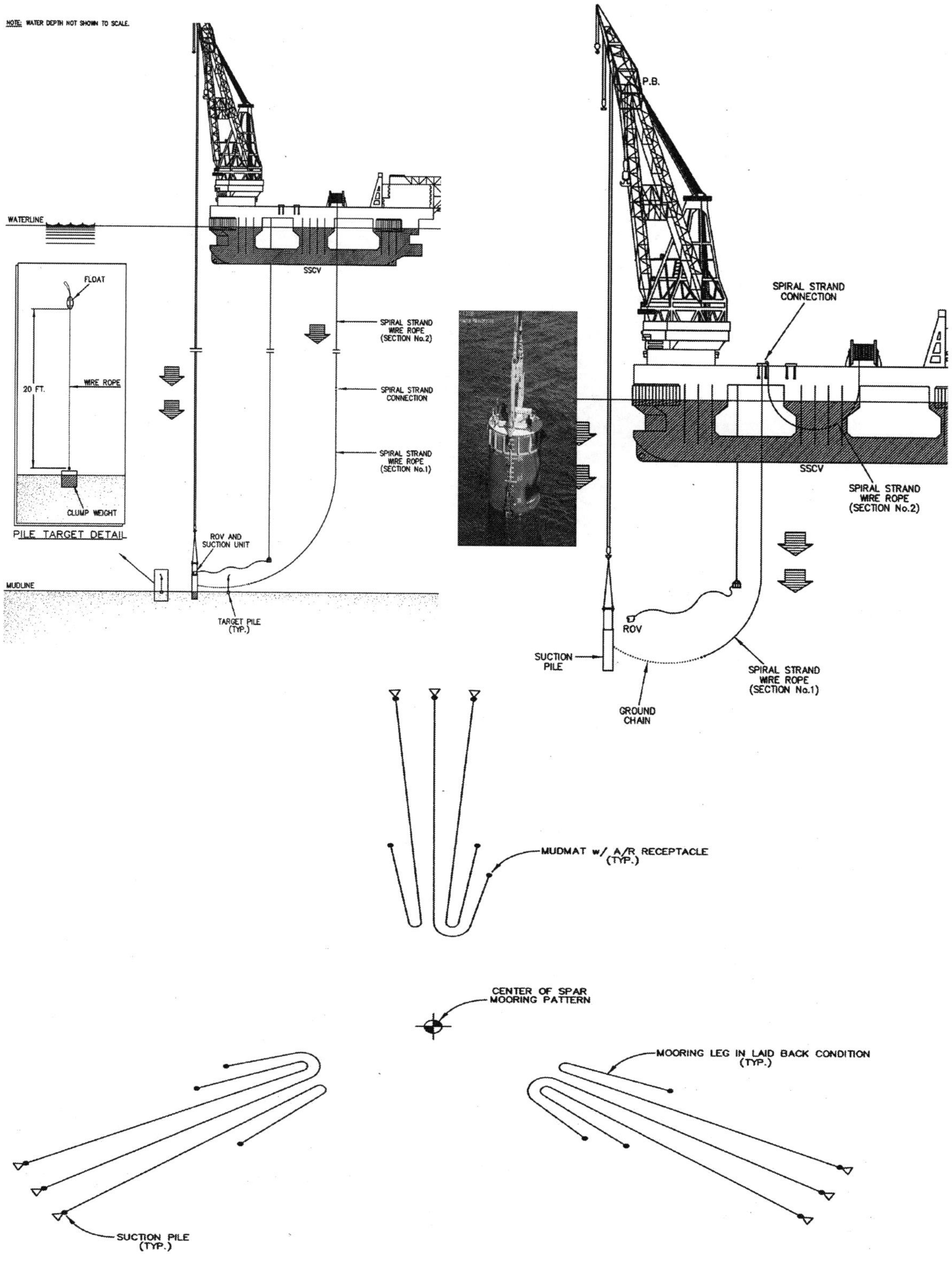

Fig. 2-18. Mooring installation

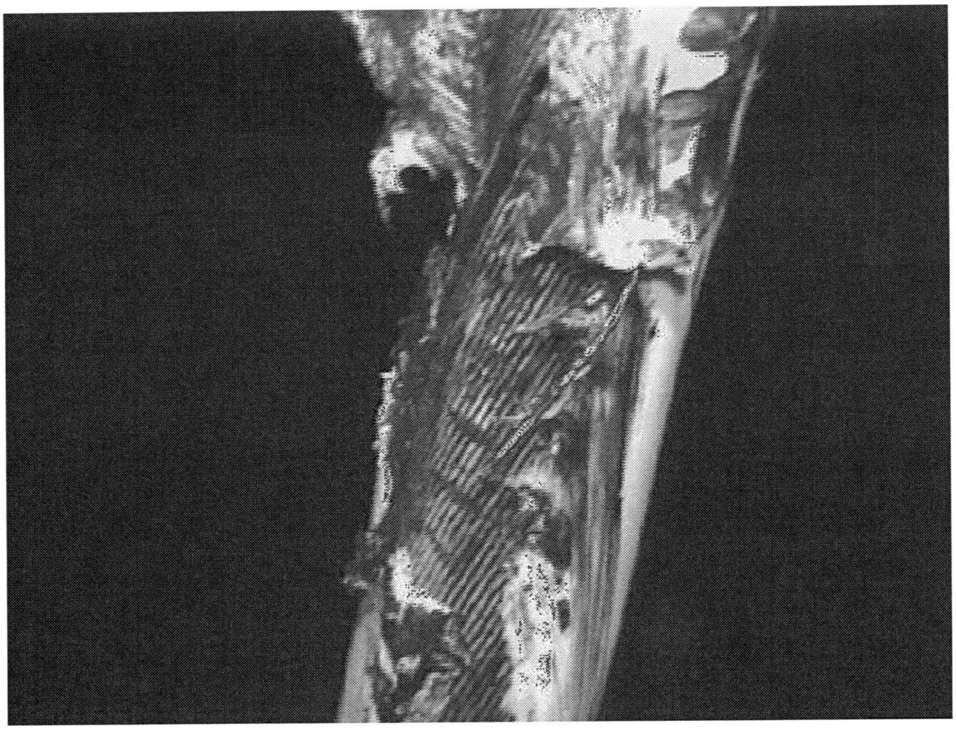

Fig. 2-19. A wire rope segment damaged during installation

Mooring Line Hookup

Mooring line hookup begins after upending of the spar hull on site. This involves picking up a pre-laid mooring line on the seabed, connecting the pull-in messenger line, transferring the mooring line tension fully to the messenger line, pulling in the messenger line through the fairlead to the stopper, and engaging the platform chain in the wildcat.

To make the spar hull storm ready as quickly as possible, a minimum number of mooring lines must be pulled in, during which tugs or other means of stationkeeping will be needed to hold and maneuver the spar hull. Once those lines are pulled in, nominal pretensions can be set.

The remaining lines can be picked up and pulled by following a similar procedure. The level of pretensions and frequency of adjustment during installation and pre-commissioning depends on details of the installation activities.

A post-installation survey should be conducted to ensure that (1) no damage is inflicted on the mooring lines and that (2) the lines are not excessively twisting beyond the manufacturer's recommended limit.

Sparing

Mooring line sparing philosophy is highly operator-dependent. However, given the long-lead time of jacketed spiral strand wire rope and its susceptibility to damage during installation, purchasing a spare segment is recommended.

Typical Installation Tolerances

- Padeye orientation (azimuth) ±7.5°
- Plumbness (verticality) ±5.0°
- Penetration - 0 ft; + 1 ft
- Position (off theoretical) < 20 ft radius

Polyester Mooring System

MMS has recently approved the application of polyester mooring system designs for BP's Mad Dog Truss spar and for Kerr-McGee's Red Hawk Cellar spar. API RP 2SM discusses the advantages of polyester mooring systems compared with steel mooring systems.

Many of the mooring system design safety factors used for steel mooring systems apply to the design of a polyester mooring system. Industry experience with designing these systems is very limited, however, despite its rapid evolution. The following section provides a brief summary of the key differences reflecting the current industry understanding. Some interesting discussions on the use of polyester moorings for MODUs can be found in the *Offshore Engineer* (November 2001).

Installation

Polyester mooring lines are prone to damage during installation. Special installation equipment is needed to avoid physical damage during handling and installation offshore.

Additionally, experience with pre-laid polyester lines indicates significant strength degradation due to ingression of sand. Therefore, polyester mooring segments must be installed at the same times as spar, preferably using a subsea mooring connector. Alternatively, the polyester segment can be pre-installed and buoyed off provided that a sufficient segment of the ground chain is lifted off the seabed at the lower end of the polyester segment to ensure minimum line tensions as well as to place it above ingression of dirt.

Nonlinear Line Properties

Polyester mooring lines have nonlinear and visco-elastic (tension rate dependent) line properties. Additionally, polyester mooring lines have significant initial bedding-in stretch and some long-term creep. The mooring analysis and design need to cover the practically expected range of variation in line tension due to stretching and creeping and in-line properties, specifically the EA. In addition, the platform chains must be long enough to accommodate the initial stretch and creep of the polyester segment(s).

Fig. 2-20. A connection between a polyester segment to chain

Rope Termination and Connection

Figure 2-20 shows the spliced termination of a polyester rope segment and connection to the splice through a purpose-designed roller using a D-shackle. The roller size is governed, among other things, by the minimum D/d ratio, for which the rope supplier should be consulted.

Insert Segment

Because long-term performance data on polyester mooring lines for permanent mooring system installations is limited at this time, insert segments must be used and later retrieved for in-place performance monitoring of the system. Both the Mad Dog and Red Hawk spar mooring system designs incorporate inserts, which are normally located immediately below the platform chain.

Procedures for removal of the insert segment should be incorporated in the mooring system design.

Minimum Line Tension

Because of the uneven load sharing between fibers in a polyester mooring line, it is recommended to maintain a minimum line tension to minimize damaging the fibers due to compression. This requirement may necessitate the use of significant ground chain per API RP 2SM. Alternatively, the system designer will need to count line tension cycles to demonstrate that the number of low-tension cycles is within the allowable limit.

In practice a polyester mooring system, which is often associated with taut-leg mooring system design, becomes a semi-taut mooring system design because of the long-ground chain required to maintain minimum line tensions and/or to avoid grounding of the polyester lines in the intact and one-line damaged condition.

References

DeLuca, Marshall (2001). Polyester is back in style., *Offshore Engineer*, November, 2001, p. 31.

Dove, P. G. S., Fulton, T. M. (1997). Why mobile drilling units will be able to moor in 10,000 ft depths, *Offshore*, Vol. 57, Issue 4, p. 32.

First Approval of Use of Synthetic Moorings as New Technology (2002), *MMS News Release*, November 7.

Garrett, D., Gordon, R., and J. Chappell. (2001). Mooring challenges increase with deepwater, rough locations, *Offshore*, Vol. 61, Issue 11, p. 72.

MMS Announces First Approval For Use of Cell SPAR as New Technology (2003), *MMS New Release*, January 30.

Sefton, Sara L., Hallam, Stuart (1999). Extending the limits of steel mooring cables for ultra-deepwater, *Offshore*, Vol. 59 Issue 5, p. 72.

Chapter 3: Hull/Mooring/Riser Coupled Spar Motion Analysis: Sensitivity against Methodological/Environmental/Empirical Parameters

By M. H. Kim, Ph.D., Professor, Civil Engineering/Ocean Research Engineering, Texas A&M University

Abstract

For nonlinear hull/mooring/riser coupled dynamic analyses of a spar, a pseudo time-domain approach is developed and used. The first-order wave-frequency and second-order difference-frequency wave loads and other hydrodynamic coefficients for the hull are calculated in the frequency domain using a second-order diffraction/radiation 3D panel program. The wave forces on the hull are then converted to time-domain signals using two-term Volterra series model. At each time step, the forces on slender members are directly calculated from Morison's formula at the instantaneous body position. The responses of hull and slender members are then simultaneously solved in a combined matrix at each time step. The case-study simulations were conducted for a classic spar designed for 6000-ft. (1829-m) water depth with 14 chain-polyester-chain mooring lines and 23 buoyancy-can-supported vertical risers in 100-year non-parallel wind, wave, and current conditions. The fully coupled hull-mooring-riser numerical results are compared with uncoupled quasi-static and semi-coupled dynamic analysis results, in which the mooring lines and risers are replaced by a set of massless springs. A series of sensitivity studies are then carried out against various analytical/environmental parameters to better understand the role of each parameter and the underlying physics. In all cases considered, the wave-frequency components remain almost the same, while the slowly varying motions change case by case. Of critical note, the equivalent static wind-force modeling may lead to significant underestimation of slowly varying surge/sway and pitch/roll responses in the absence of dynamic wind loading. The importance of the effects of buoyancy cans on spar pitch motions is also addressed.

Introduction

Since the first installation of the Oryx Neptune spar in the Gulf of Mexico in 1996, floating spars have become increasingly popular as an economic and reliable oil production concept in deep or ultra-deep water, particularly in the remote areas without pipeline infrastructure. So far, two competitive spar concepts have mainly been used (Halkyard 1996)—classic spars using a deep-draft hollow vertical cylinder and truss spars using a combination of relatively shallow-draft hollow cylinder and truss structure extended below it. Recently, a new spar concept called cell spar was also proposed and designed. When compared to the classic spar, both truss and cell spars are more transparent against waves and currents and, thus, known to be less sensitive to vortex induced motion (VIM).

As water depth increases, the portion of the mooring and riser mass becomes larger compared to the hull mass, and the resulting inertia/loading and damping effects are expected to be important. A reliable program to solve the hull/mooring/riser coupled dynamics as an integrated system is required to accurately simulate the inertia/loading and damping effects of the mooring lines and risers on the hull motions needs. The coupled analysis can in principle be done both in time and frequency domain (Ran 2000). The frequency-domain approach is computationally more efficient, but the nonlinear drag forces have to be linearized. Conversely, the time-domain approach is computationally more intensive, but the nonlinear drag forces and other nonlinear and transient effects can be preserved.

For some systems, the coupling effects may magnify the extreme hull responses; however for platforms in deep waters, the coupling effects more likely lead to smaller extreme responses due to additional damping from risers and mooring lines. Thus, the coupled analysis more likely results in less expensive mooring/riser system and more cost-effective design.

Until recently, the design of risers and mooring lines has mostly been based on either uncoupled quasi-static analysis (model mooring and risers as massless linear or nonlinear spring, calculate hull responses, and estimate the mooring tension from static-offset curve) or semi-coupled dynamic analysis (model mooring and risers as massless linear or nonlinear spring, calculate hull responses, input the calculated motions at the fairlead, and run line dynamics program for each mooring line). In the absence of more accurate time-domain hull/mooring/riser coupled analysis tools, the reliability of those approximation methods has primarily been checked against model test data. However, as water depth increases significantly, the length of mooring lines and risers cannot be correctly scaled due to the depth limitation of existing wave basins. Under this circumstance, the statically-tuned but dynamically-distorted truncated mooring-riser system had to be employed, which means that the dynamic coupling effects between hull and slender members may not be well reproduced in the model test.

As a result, offshore industry has paid a lot of attention to the development and verification of reliable numerical analysis tools that can accurately simulate the hull/mooring/riser coupled dynamics of deepwater platforms. Once fully verified and calibrated with the model test data with truncated riser and mooring lines, the coupled-analysis computer program can be used to extrapolate to larger depths accommodating the full length of risers and mooring lines. This approach, called the experimental-numerical hybrid model-testing method, partly solves the problem related to the unavailability of deepwater model-test basins. Therefore, a reliable computer program for coupled dynamic analysis is critical for ultra-deepwater projects.

To evaluate the performance/variability of existing time-domain hull/mooring/riser coupled dynamic analysis tools, a special project was launched in 1999 by the Deep-Star Offshore Industry Consortium to collect and compare the various computational results produced by major oil companies, research institutes, wave basins, and offshore contractors. For the project, three different deepwater platforms—spar, TLP, and FPSO—designed for 3000-ft., 6000-ft., and 10,000-ft. water depths were considered. All the design parameters and environmental conditions were given, and many contributors volunteered to analyze either one or two structures using their own computer programs and procedures. The numerical results produced were later compared with the experimental results conducted in the MARIN deepwater model basin. The result of this comparison study is summarized in Colby et al. (2000), Ma et al. (2000), Gupta et al. (2000), Wichers et al. (2001), and Steen et al (2004). From this study, it was found that the hull/mooring/riser coupled dynamic analysis can reasonably predict the global spar responses in the given survival condition (100-year storm). In case of a loop-current dominant environment, the VIM of spar hull plays an important role and, therefore, should be included through reliable numerical modeling. In this case, it is more difficult to compare between numerical and experimental results due to the inherent turbulence of basin-generated currents.

The calibration of numerical models can be done by varying empirical parameters, such as the drag coefficients of hull and lines. The viscous effects on hull drift motions also must be calibrated. To better understand the underlying physics and the role of relevant parameters, the study includes the sensitivity of spar global motions and mooring tension by varying empirical and environmental parameters or analysis methodologies. The spar designed for 6000-ft. water depth in the DeepStar study was selected for this sensitivity study. It has 14 taut chain-polyester-chain mooring lines and 23 buoyancy-can-supported vertical risers. The sensitivity study results may be useful in developing an efficient calibration scheme against full-scale or model-testing data. Several important and useful conclusions are drawn based on the present sensitivity study and are summarized in the last section.

Specification of Classic Spar (6000 ft.)

To illustrate the use of hull/mooring/riser coupled-dynamic-analysis tool for spar global responses and the corresponding line tensions, the following classic spar designed for 6000-ft. water depth is selected.

Production Level	55,000 bpd of oil and 72 mmscfd of gas

- Hull

Displacement	53,600 m.ton
Total displacement	220,740 m.ton
Diameter	122 ft. (37.2 m)
Length	705 ft. (214.9 m)
Draft	650 ft. (198.1 m)
Hard tank depth	220 ft. (67 m)
Well bay dimensions (25 slots)	58 ft. × 58 ft. (17.7 m × 17.7 m)
KB	540 ft. (164.6 m)
KG	462 ft. (140.8 m)
KG (based on total displacement)	314 ft. (95.7 m)
Radius of gyration (based on total displacement)	pitch=221 ft. (67.4 m) yaw=28.5 ft. (8.7 m)
Drag force coefficient	1.15
Wind force coefficients	0.0558 (kips/(ft./sec)^2)
Center of pressure	722 ft. (220 m) ABL (above base line)

In this spar design, there are 18 production risers, one drilling riser, two water injection risers, and two oil/gas export risers. The 23 risers are arranged in a very compact manner as shown in the Figure 3-1. The weights of the risers are supported by buoyancy cans at the top.

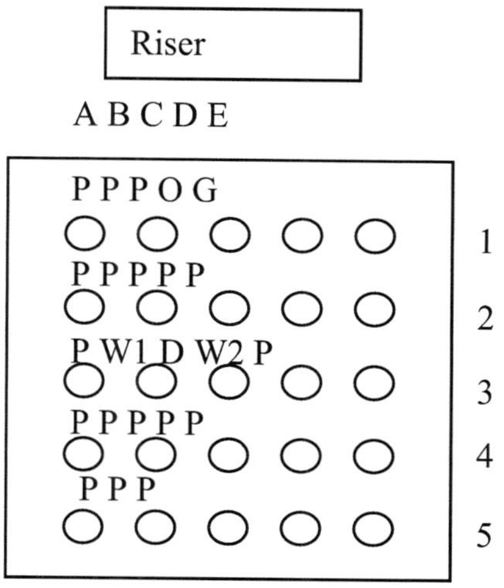

Fig. 3-1. Compact arrangement of risers

Notes:

1. Displacements or net buoyancy includes hull, topsides, mooring, and hard tanks.
2. Total displacement includes hull, topsides, mooring, hard tanks, and water in soft tanks and moonpool.
3. For simplicity, all risers below are assumed to be top-tensioned, near-vertical risers supported by buoyancy cans. The riser top tensions given below are to be taken at the keel level of the spar. The risers are restrained from lateral motions at the keel. (This assumption is introduced to avoid complicated riser modeling inside the spar moonpool. The adequacy of this simple modeling will be further discussed in the following section.)

- **Risers** (1 kip = 4448 Newton, 1 m=3.281 ft.)

Table 3-1. Riser Specification

Riser	Top tension (kips)	Outer diameter (in.)	AE (kips)	Weight (dry / wet) (lb./ft.)
Drilling	1470	21	2.70E6	400 / 245
Production 1~18	786	14.25	6.73E5	202 / 131
Water injection 1	261	8.625	4.13E5	69.4 / 43.5
Water injection 2	151	6.625	2.44E5	40.4 / 25.1
Oil export	660	16	1.04E6	199 / 110
Gas export	306	16	1.04E6	140 / 51
Note: All added mass and drag coefficients are 1.0.				

- **Mooring System**

Mooring type	Chain-polyester-chain taut
Mooring pattern	14–point, taut-leg omni-directional spread
Mooring line composition	
Platform section	300 ft. ×4.625 in. K4 studless chain
Riser section	7800 ft. ×8.27 in. polyester
Ground section	400 ft. ×4.625 in. K4 studless chain
Scope ratio (length/depth)	1.41
Fairlead location	300 ft. ABL (above base line)
Pretension	530 kips
Current force coefficient of chain	2.45
Current force coefficient of wire	1.2

Table 3-2. Mooring Line Specifications

Mooring Lines	Dry/Wet weight (lb./ft.)	MBL (kips)	AE (kips)	Added mass (lb./ft.)
4.625 in. K4 studless chain	193.44/ 168.29	2653	231,900	25.15
8.27 in. Polyester	24.55/ 5.22	2876	71,600	19.33

Assumptions and Simplifications

In the present study, all risers are in default truncated at the keel, which excludes the additional pitch restoring moments by buoyancy cans. The effects of buoyancy cans on spar pitch motions are also investigated and discussed.

It is also assumed that there is no spiral strake on the cylindrical hull for simplicity. The spiral strake generally increases hull drag coefficients both in surge, pitch, and heave. It can also slightly increase the added mass and wave exciting force on the hull, particularly in heave mode, and/or cause heave-yaw coupling through screw-like effects.

The vortex induced vibrations (VIVs) of spar hull and risers/mooring are not included in the present numerical study. The hull VIV may cause appreciable figure 8-shaped or banana-shaped periodic oscillations in the transverse plane (normal to the current/wave direction). The vortex shedding frequency can roughly be estimated from the Strouhal number $Sr = Df / V$ for the given condition, where V is stream velocity, f=vortex shedding frequency, and D=cylinder diameter. The VIV of risers and mooring lines tend to increase the in-line drag forces. The spar hull VIV is particularly important in the presence of strong and deeply penetrating loop current.

Slowly varying wave drift damping is neglected in the present analysis. The wave drift damping can be obtained numerically from the rate of change of mean drift force with respect to the forward velocity U at U=0 (Emmerhoff & Sclavounos 1992, Kim & Kim 1995). Alternatively, it can be obtained from the gradient of mean drift forces, which is called Aranha's formula (Clark et al. 1993). The wave drift damping is proportional to the square of the wave amplitude, so it becomes more important in higher waves. In the case of spar motion analysis, the effects of wave drift damping are in general very small (Kim 2003).

Wave and current interactions are neglected. The wave forces and hydrodynamic coefficients can change in the presence of currents (Faltinsen 1990, 1994; Kim and Kim 1995). The simplest correction is the modification due to the change of wave frequencies to encounter frequencies. In the present example, these effects are expected to be small and not included. As a matter of fact, currents play a more important role in changing drag forces through Morison's equation, which is considered here.

In this study, only long-crested, unidirectional irregular waves are considered. However, the assumption of uni-directionality does not necessarily warrant conservative prediction even in in-line responses. In other words, the case with directional spreading may produce larger maximum slowly varying wave loading and body responses. Kim (1992) reported that the second-order difference-frequency wave loading can be significantly increased when two waves are approaching from different angles. This situation may happen in crossed sea conditions—storm-generated waves from one direction and swell from another direction. The possible increase of low-frequency motions due to directional spreading is beyond the scope of the present study.

The water inside the moonpool was included as part of body mass. The effects of water jets through the gaps of risers and riser-guide holes should be negligible. It is also assumed that the sloshing motions of the water inside the moonpool do not influence the global motions of the spar platform. The additional free surface of the moonpool may slightly alter the pitch restoring moment and natural frequency, which can be estimated and incorporated into the pitch-roll restoring coefficient.

In the mooring calculations, it is assumed that the given polyester lines satisfy the linear stress-strain relationship. In reality, the stress-strain relationship of polyester lines may deviate appreciably from the linear one depending on the material, number of fabrics, and how they are intertwined and fabricated. A more complicated mooring analysis allowing larger elongation and nonlinear stress-strain curves is introduced in Arcandra (2001).

If Morison's formula is used instead of diffraction theory (this may be possible only for survival condition), the wave forces on the hull can be calculated at each time step for instantaneous body position. In this case, a more rigorous form of Morison formula (Rainey 1989; Eatock Taylor et al. 1992; Kim & Chen 1994; Mekha et al. 1996; Cao & Zhang 1996) is recommended. The mean wave drift force must be separately estimated from the potential theory. Wave kinematics above mean water level (MWL) requires a proper stretching method to be employed because the direct use of linear solutions (exponential growth) may lead to significant over-prediction. In the present study, the uniform extrapolation method, in which the kinematics above MWL are replaced by those at MWL, is adopted.

When second-order diffraction theory is used, the effects of body motions and fluctuating free surface are already included. When the platform undergoes large amplitude slow drift motions, a phase correction of e^{ikx} (x=instantaneous offset position of platform) can be used to roughly account for its effect.

The Coulomb friction between risers and riser guides is not considered here, which is expected to decrease the heave motion (Koo 2003).

Numerical Modeling of Spar Hull, Mooring, and Risers

For more rigorous hull-mooring-riser coupled dynamics, the fully nonlinear wave forces at the instantaneous hull position need to calculated at each time step. However, the fully nonlinear free-surface and body interaction problem is extremely expensive and not likely to be practical for the coming years (Celebi et al. 1998, Kim 1995). As a practical alternative, a pseudo time-domain approach is employed in the present study. The first- and second-order wave forces are first calculated in the frequency domain then the wave forces are converted to the time domain. The first- and second-order wave forces as well as added mass and hydrodynamic damping are calculated from the second-order diffraction/radiation 3D panel program WAMIT (Korsmeyer et al. 1988, Lee et al. 1991). For spar motion analysis, the sum-frequency wave forces are not important and, thus, are not included.

The wave-force time histories were generated in the time domain using two-term Volterra series expansion (Kim and Yue 1991, Ran and Kim 1996). The frequency-dependent radiation damping was included in the form of convolution integral in the time domain simulation. In the second-order diffraction/radiation calculation, both body surface and free surface must be discretized. The spar hull is discretized by 1504 panels and the free surface is discretized by 704 elements inside a truncation radius of 427 ft. (130 m), as shown in Figure 3-2. The discretization was shown to be satisfactory when sum-frequency components are not considered.

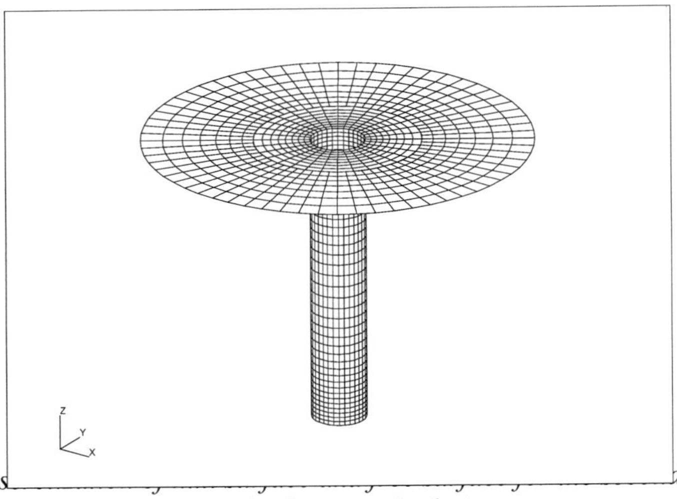

Fig. 3.2. Des[illegible]*d-order diffraction/radiation calculation*

The hull/mooring/riser coupled statics/dynamics are calculated by the computer program WINPOST (Ran and Kim 1996, Kim et al. 1999). The mooring lines are hinged at the fairlead and the anchor. The risers are held vertically at the spar keel by buoyancy cans. Therefore, the riser tension is not included in the vertical static equilibrium of the system. When risers are supported by hydraulic or pneumatic tensioner-like system, the riser tension should be included in the static equilibrium in vertical plane. The calculated platform mass for the given condition is 2.17×10^8 kg.

The hull is positioned by 14 taut-leg, omni-spread chain-polyester-chain mooring lines, and 23 air-can-supported vertical risers. The total riser top tension is 16,996 kips (7.56×10^7 N). Each mooring line and riser is modeled by 12 high-order elements, totaling 444 high-order elements for mooring and risers. The theoretical background of the employed finite element method (FEM) is explained in the next section. The convergence of the grid of mooring lines and risers was checked by doubling the number of elements for selected lines. Figure 3-3 shows the arrangement of hull, mooring lines, and risers with distorted scale.

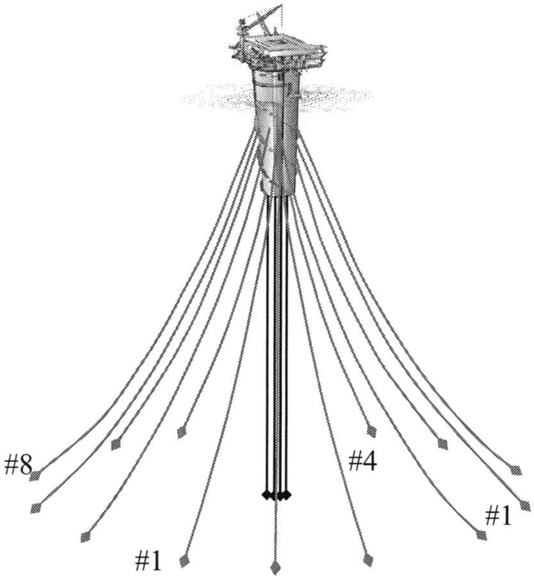

Fig. 3-3. The arrangement of mooring lines and risers (with distorted scale)

- *Static-Offset Tests (in calm water without current)*

The platform stiffness can be determined from numerical static offset test. The surge static offset test was conducted by pulling vertical center of gravity (VCG) in the horizontal direction in calm water. Typical results for surge offsets are shown in Figure 3-4.

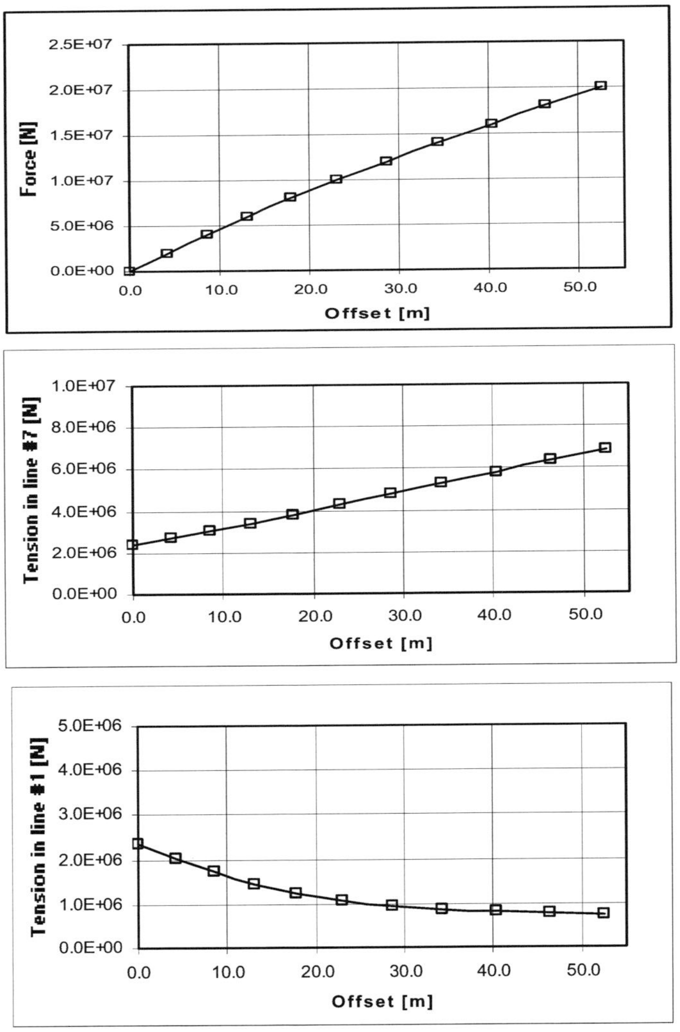

Fig. 3-4. Static offset test and line tension for surge

The surge static-offset test for the present chain-polyester-chain spread taut mooring system shows a weakening trend, while most steel-wire/chain system shows hardening characteristics. The calculated surge stiffness in the 10 to 40 m range is 5.31 to 4.31e+5 N/m. All the mooring line tensions at the respective offset positions are separately checked. Two of such examples are shown for taut line #7 and slack line #1, as in Figure 3-4b-c.

- *Free-Decay Tests (in calm water without current)*

The system's natural periods and damping can be determined from free-decay tests. For deepwater moored platforms, the damping from mooring and risers can be of particular importance, and they can be included by using hull-line coupled analysis.

Table 3-3. Surge and Heave Natural Periods from Free-Decay Tests

	Surge	Heave
6000-ft. spar (w/o risers)	185s	28s
6000-ft. spar (w risers)	191s	28s

Table 3-4. Damping From Free-Decay Tests Estimated From the First Four Peaks Assuming Linear Damping

(The number in parenthesis is the amplitude range)

	Surge	Heave
6000-ft. spar (without risers)	4.6-3.3% (10-5 m)	0.6% (2-1.8 m)
6000-ft. spar (with risers)	6.2-4.3% (9-3 m)	0.6% (2-1.8 m)

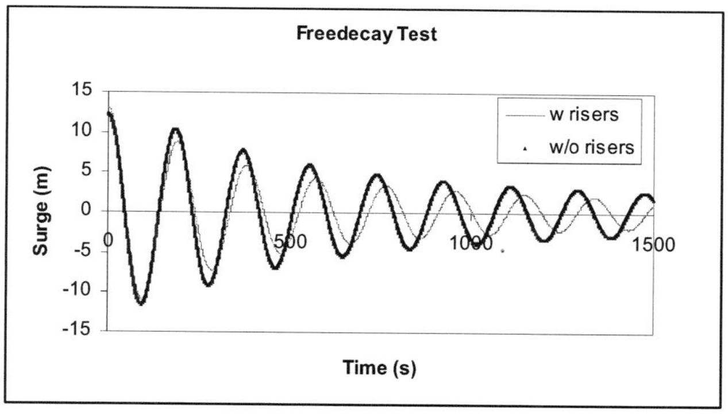

Fig. 3-5. Free-decay test results for surge with and without risers

To see the effects of risers (mostly the amount of damping from risers) in the free-decay tests, a simpler riser model was developed, that is, all the 23 risers are replaced by a single equivalent *massless* riser at the center with the same total tension 16,996 kips (12.6×10^7 N). The resulting surge/sway stiffness of the riser at the keel can approximately be calculated from the total tension divided by length. Figure 3-5 shows a numerical free-decay test result for surge mode. Tables 3-3 and 3-4 show the natural periods and damping for surge and heave from respective numerical free-decay tests.

In most cases, the damping increases for larger oscillation amplitudes. With the simplified massless riser model, the surge damping is appreciably reduced, while the heave damping remains the same. Similarly, there is no difference in heave natural periods, but the surge natural period is increased after including the riser mass and drag. From this result, it is clear that the riser mass and drag must be correctly modeled for more accurate response results, particularly when strong currents exist.

Fundamentals of Motion Analysis

Linearized Equation of Motion in Time and Frequency Domain.

As explained in the preceding section, wave forces, added mass, and radiation damping were first calculated in the frequency domain using a 3D diffraction/radiation panel program WAMIT. In the context of linear theory, the motion of a floating body in a regular wave can be obtained from

$$([m]+[m_{aij}])\ddot{\tilde{\xi}}^{(1)} + ([b_{ij}]+[B_{ij}])\dot{\tilde{\xi}}^{(1)} + ([c_{ij}]+[K_{ij}])\tilde{\xi}^{(1)} = \tilde{F}^{(1)} \tag{3-1}$$

where 6DOF motions $\tilde{\xi} = (\Xi_x, \Xi_y, \Xi_z, \Omega_x, \Omega_y, \Omega_z)^T$, $[m], [m_{aij}], [b_{ij}]$, and $[c_{ij}]$ are mass, added mass, radiation damping, and hydrostatic restoring coefficient matrices, respectively. The symbol $[B_{ij}]$ is the linearized equivalent viscous damping from both hull and slender members and $[K_{ij}]$ is the linearized equivalent mooring-line/riser stiffness. The mass matrix is given by

$$\begin{pmatrix} m & 0 & 0 & 0 & mz_g & -my_g \\ 0 & m & 0 & -mz_g & 0 & mx_g \\ 0 & 0 & m & my_g & -mx_g & 0 \\ 0 & -mz_g & my_g & I_{xx} & I_{xy} & I_{xz} \\ mz_g & 0 & -mx_g & I_{yx} & I_{yy} & I_{yz} \\ -my_g & mx_g & 0 & I_{zx} & I_{zy} & I_{zz} \end{pmatrix} \tag{3-2}$$

where m = body mass, (x_g, y_g, z_g) = coordinate of the center of gravity, and I = mass moment of inertia. The linearized viscous damping depends on the unknown motion amplitudes; therefore, an iterative procedure has to be used. For moored structures especially in deep water, a major part of B may come from risers and mooring lines (mooring-line damping), as was pointed out by Webster (1995).

The hydrostatic restoring coefficients are dependent only on hull-geometry and can be calculated from the following formulas:

$$c_{33} = \rho g \iint_{S_B} n_z dS, \; c_{34} = \rho g \iint_{S_B} y n_z dS, \; c_{35} = -\rho g \iint_{S_B} x n_z dS,$$

$$c_{44} = \rho g \iint_{S_B} y^2 n_z dS + \forall \rho g z_b - mgz_g, \; c_{45} = -\rho g \iint_{S_B} xy n_z dS,$$

$$c_{55} = \rho g \iint_{S_B} x^2 n_z dS + \forall \rho g z_b - mgz_g, \; c_{46} = -\rho g \forall x_b + mgx_g,$$

$$c_{56} = -\rho g \forall y_b + mgy_g \tag{3-3}$$

where $c_{ij} = c_{ji}$ for all i, j in the above except for c_{46}, c_{56} and for all other values of indices $c_{ij} = 0$. The symbols $\forall$ and (x_b, y_b, z_b) represent the displaced volume and the coordinate of the center of buoyancy of a body.

When the natural frequencies of a system are much smaller (or larger) than predominant wave frequencies and the system damping is small near the resonance region, the slowly-varying (or high-frequency) responses excited by the second-order difference-frequency (or sum-frequency) wave forces can be large. In such a case, it is important to include second-order difference-frequency (or sum-frequency) wave loads in the global motion analysis.

When pseudo nonlinear time-domain approach is used, eq 3-1 can be rewritten in the following form:

$$([m]+[m_{aij}(\infty)])\ddot{\tilde{\xi}} + \int_0^{\infty} \tilde{R}_{ij}(t-\tau)\dot{\tilde{\xi}}(\tau)d\tau + ([c_{ij}]+[K_{ij}(\tilde{\xi})])\tilde{\xi} = \tilde{F}(t) \qquad (3\text{-}4)$$

where $\tilde{F}(t)$ are total excitation forces from environment and mooring/risers. The convolution integral accounts for the contribution from frequency-dependent radiation damping. The retardation function $\tilde{R}$ is

$$\tilde{R}(t) = \frac{1}{2\pi}\int b_{ij}(\omega)\cos\omega t d\omega \qquad (3\text{-}5)$$

In random waves, the first- and second-order wave loads can be expressed as follows using the two-term Volterra series expansion:

$$F^{(1)}(t) = \mathrm{Re}\sum_{i=1}^{N} A_i f_i e^{-i\omega_i t} \qquad (3\text{-}6)$$

$$F^{(2)}(t) = \mathrm{Re}\sum_{i=1}^{N}\sum_{j=1}^{N} \{A_i A_j^* f_{ij}^-(\omega_i,\omega_j)\, e^{-i(\omega_i-\omega_j)t} + A_i A_j f_{ij}^+(\omega_i,\omega_j) e^{-i(\omega_i+\omega_j)t}\} \qquad (3\text{-}7)$$

where A is complex wave amplitude, N is the number of wave components, and f_i and $f_{ij}^{\pm}$ are linear (LTF) and quadratic force transfer functions (QTF), that is, respective wave forces due to unit-amplitude monochromatic and bichromatic incident waves. When the natural frequencies of a moored platform are very small, it is reasonable to approximate f_{ij}^- by its diagonal component f_{ii}^- so that they can be more easily calculated (Newman 1974, Kim and Yue 1989a). The method is called Newman's approximation. The wave drift damping in random waves can also be simulated in the same manner as in eq 3-7. The detailed spectral and statistical analyses of the two-term Volterra model are summarized, for example, in Kim & Yue (1989a, 1991).

Nonlinear Hull/Mooring/Riser Coupled Dynamic Analysis in Time Domain

When water depth is large, hull/mooring/riser coupling effects are expected to be appreciable, and their dynamics should be solved simultaneously as an integrated system. There are two approaches to coupled analysis—the iteration method and the combined matrix method. The former uses iteration between hull and line dynamics during time marching until convergence is reached; the latter solves the hull and lines as an integrated system in a combined matrix.

The static/dynamic analysis of mooring lines and risers uses an extension of the theory developed for long slender rods by Garrett (1982). Assuming that there is no torque or twisting moment, a linear momentum conservation equation can be derived with respect to a position vector $\vec{r}(s,t)$, which is a function of arc length s and time t:

$$-(B\vec{r}'')''+(\lambda\vec{r}')'+\vec{q} = m\ddot{\vec{r}} \tag{3-8}$$

$$\lambda = T - B\kappa^2 \tag{3-9}$$

where primes and dots denote spatial s-derivative and time derivative, respectively, B (=EI; E=Young's modulus; I=sectional moment of inertia) is the bending stiffness, T the local effective tension, κ the local curvature, m the mass per unit length, and $\vec{q}$ the distributed force on the rod per unit length. The scalar variable λ can be regarded as a Lagrange multiplier. The rod is assumed to be elastic and extensible, thus the following condition is applied (Pauling et al. 1986).

$$\frac{1}{2}(\vec{r}\cdot\vec{r}-1) = \frac{T}{A_t E} \approx \frac{\lambda}{A_t E} \tag{3-10}$$

where, $A_t = A_e - A_i$ (= outer – inner cross sectional area). For these equations, geometric non-linearity is fully considered and no special assumption is necessary concerning the shape or orientation of lines. The benefit of this equation is that eq 3-8 is directly defined in the global coordinate system and does not require any transformations to the local coordinate system, which saves overall computational time significantly.

The normal component of the distributed external force on the line per unit length, q_n, is given by a generalized Morison equation:

$$q_n = C_I \rho A_e \dot{v}_n + C_D \frac{1}{2}\rho D|v_{nr}|v_{nr} + C_m \rho A_e \ddot{r}_n \tag{3-11}$$

where C_I, C_D and C_m are inertia, drag, and added mass coefficients, and $\dot{v}_n, v_{nr}$, and $\ddot{r}_n$ are normal fluid acceleration, normal relative velocity, and normal structure acceleration, respectively. The symbols ρ and D are fluid density and local diameter. In addition, the effective weight, or net buoyancy, of the rod should be included in q_n as a static load.

A finite element method similar to Garrett (1982) has been developed to solve the above mooring dynamics problem and the details of the methodology are given in Ran et al. (1997) and Ran (2000). The FEM allows any combination of mooring types and materials as long as their deformations are small and within proportional limit.

The upper ends of the mooring lines and risers are connected to the hull fairlead through generalized elastic springs and dampers. The combination of linear and torsional springs can model arbitrary connection conditions. The forces and moments proportional to the relative displacements are transmitted to the hull through the connection points. The transmitted forces from mooring lines and risers to the platform are given by

$$\tilde{F}_P = \tilde{K}(\tilde{T}\tilde{u}_P - \tilde{u}_I) + \tilde{C}(\tilde{T}\dot{\tilde{u}}_P - \dot{\tilde{u}}_I) \tag{3-12}$$

where $\tilde{K}, \tilde{C}$ are stiffness and damping matrices of connectors at the connection point, and $\tilde{T}$ represents a transformation matrix between the platform origin and connection point. The symbols $\tilde{u}_P, \tilde{u}_I$ represent column matrices for the displacements of the platform and connection point.

Then, the following hull response equation can be combined into the riser/mooring-line equations in the time domain:

$$\begin{aligned}(\tilde{M} + \tilde{M}_a(\infty))\ddot{\tilde{u}}_p + \int_0^{\infty} \tilde{R}(t-\tau)\dot{\tilde{u}}_p d\tau + \tilde{K}_H \tilde{u}_p = \\ \tilde{F}_D + \tilde{F}^{(1)} + \tilde{F}^{(2)} + \tilde{F}_p + \tilde{F}_w + \tilde{F}_c + \tilde{F}_{WD}\end{aligned} \tag{3-13}$$

where $\tilde{M}, \tilde{M}_a(\infty)$ are the matrices of body mass including moonpool water and added mass at infinite frequency, $\tilde{R}$ = retardation function (inverse cosine Fourier transform of radiation damping) matrix, $\tilde{K}_H$ = hydrostatic restoring coefficient, $\tilde{F}_D$ = drag force matrix on the hull, $\tilde{F}^{(1)}, \tilde{F}^{(2)}$ = first- and second-order wave load matrix on the hull, $\tilde{F}_p$ = transmitted force matrix from risers and mooring lines through the interface, $\tilde{F}_w$ = dynamic wind loading, $\tilde{F}_c$ = current loading on hull, and $\tilde{F}_{WD}$ = wave drift damping force matrix. To check the relative importance of $\tilde{F}_{WD}$, the diagonal components of the wave-drift-damping matrix were calculated in the frequency domain by using the mean-drift-gradient method suggested by Aranha. The time series of wave drift damping can then be generated based on Newman's (diagonal) approximation. In the present spar study, however, the wave drift damping is found to be small compared to hull viscous damping and mooring/riser damping and, thus, was not included.

The added mass at infinite frequency was obtained from Kramers-Kronig relation. For the time series of $\tilde{F}^{(1)}, \tilde{F}^{(2)}$, and $\tilde{F}_{WD}$, a two-term Volterra series was used (Kim and Yue 1991). The hull drag force in the normal direction was calculated with respect to the instantaneous hull position based on the Morrison drag formula with relative velocity squared. In the present study, the wave kinematics above MWL were approximated by those at z = 0, that is, uniform extrapolation.

The 6 × 6 rigid-body hull motions are combined into the banded matrix of FEM for slender members. The combined matrix is solved simultaneously for each time step. Therefore, the combined matrix method can handle the full hull-riser-mooring coupling without the iteration procedure. Most existing hull/mooring/riser coupled dynamics programs are based on the iteration method between mooring line and hull dynamics, which is typically simpler in program development and implementation but may not warrant convergence. In addition, the iteration method cannot handle the full hydrodynamic interactions among multiple bodies.

The static problem of the integrated system was solved using Newton's iterative method. The dynamic problem was integrated using an efficient and reliable time-marching scheme similar to Adams-Moulton method (Garrett 1982). In the dynamic program, special consideration is required because the time derivatives of λ do not appear in the equations and the added mass matrix is a function of the instantaneous position. In addition, the free-surface fluctuation and possible contact of mooring lines and catenary risers with the seafloor require special consideration.

Environmental Condition

A typical 100-year hurricane in the Gulf of Mexico with significant wave height of 12.2 meters and peak wave period of 14 seconds is selected as the wave environment. As for wind, the one-hour mean wind speed (at 10 m height) of 41.1 m/s is used, and the time dependent wind velocity is generated from the corresponding API wind spectrum. The wind direction is assumed to be 30° left of waves. As for currents, a storm-driven shear current is assumed. The current is assumed to flow from 30° right of wave direction. The non-collinear environmental condition is summarized in Table 3-3.

The JONSWAP spectrum used here is the same as that of Hasselman *et al.* (1973) with enhancement parameter γ=2.5:

$$S(\omega) = \alpha g^2 \omega^{-5} \exp\left[-1.25\left(\frac{\omega}{\omega_0}\right)^{-4}\right] \gamma^{\exp\left[-\frac{(\omega-\omega_0)}{2\tau^2\omega_0^2}\right]} \tag{3-14}$$

where γ is the peakedness parameter, and τ is the shape parameter (0.07 for $\omega \le \omega_0$ and 0.09 for $\omega \ge \omega_0$). The value of α is related to a prevailing wind velocity of U_w and a fetch of X, and can be written as

$$\alpha = 0.076(X)^{-0.22} \tag{3-15}$$

The shape of the JONSWAP spectrum used for the present study is presented in Figure 3-6.

Table 3-3. Design Environmental Condition

Designation	Unit	Value
Waves		
H_s	*m*	12.2
T_p	*sec.*	14
Wave spectrum	JONSWAP (γ=2.5)	
Wave direction	*deg.*	180 (to West)
Wind		
Wind speed (1-hour)	*m/s*	41.1 @ 10 *m*
Wind spectrum	API RP 2A-WSD	
Wind direction	*deg.*	210
Storm current profile (linear between points)		
Depth: 0 *m*	*m/s*	1.07
: 60.96 *m*	*m/s*	1.07
: 91.44 *m*	*m/s*	0.09
: seabed	*m/s*	0.09
Current direction	*deg.*	150

The 1-hour wind speed used for the API wind spectrum is based on the recurrence period of 100 years. The API wind spectrum has the following expression.

$$S(\omega) = \frac{\sigma^2(z)}{2\pi f_p \left[1 + \frac{1.5\omega}{2\pi f_p}\right]^{5/3}} \tag{3-16}$$

where f_p is the average factor derived from measured spectrum and is given by

$$f_p = \frac{0.025 V_w(z)}{z} \tag{3-17}$$

The symbol $\sigma(z)$ is the standard deviation of wind speed and related to turbulence intensity. The value of $\sigma(z)$ can be expressed as

$$\sigma(z) = 0.15\left(\frac{z}{20}\right)^{-0.125} V_w(z) \tag{3-18}$$

where $V_w(z)$ is the 1-hour mean wind speed (*m/s*) *z* meters above water level. The corresponding wind velocity spectrum used in the present study is plotted in Figure 3-7.

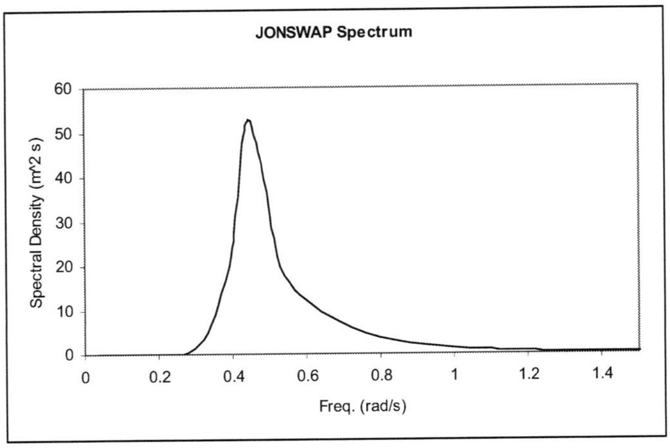

Fig. 3-6. JONSWAP wave spectrum

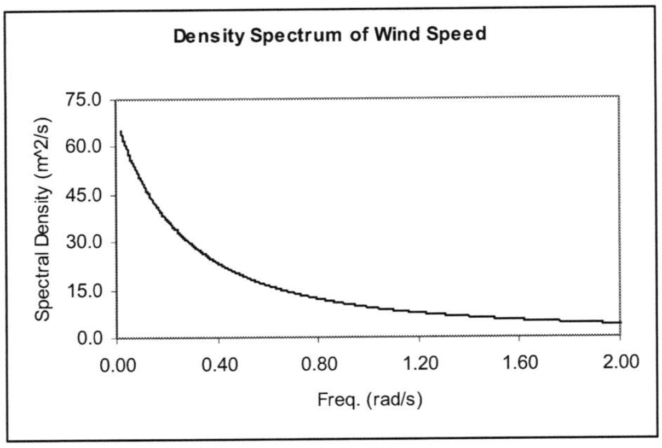

Fig. 3-7. API wind velocity spectrum

Time-Domain Simulation Results

So far, the main features and theoretical background of the hull/mooring/riser coupled dynamic analysis program are explained. In this section, we present a typical simulation result for spar global responses under 100-year storm condition.

The drag coefficients used for the present wave-force calculation are 1.0 for risers, 1.2 for polyester, 2.45 for chain, and 1.15 for spar hull. The inertia coefficient for all slender members was 2. The wave kinematics were generated at each time step and used in the Morison-force calculations for slender members.

The low- and wave-frequency regions are defined as 0-0.2, and 0.2- 1.2 (rad/s), respectively. The statistics of the fully-coupled, time-domain simulation results are summarized in Tables 3-5 and 3-4. The time step used in the time-domain simulation is 0.065 seconds, and the total simulation time is 3 hours. The time histories of surge, heave, pitch, and #2-line (highest-tension member) tension are shown in Figure 3-6. The environmental forces are gradually applied to the platform using a ramp function at the start of the time-domain simulations. The ramping period of 500s was not included in the statistical calculations. All the hull responses presented are with respect to the center of gravity.

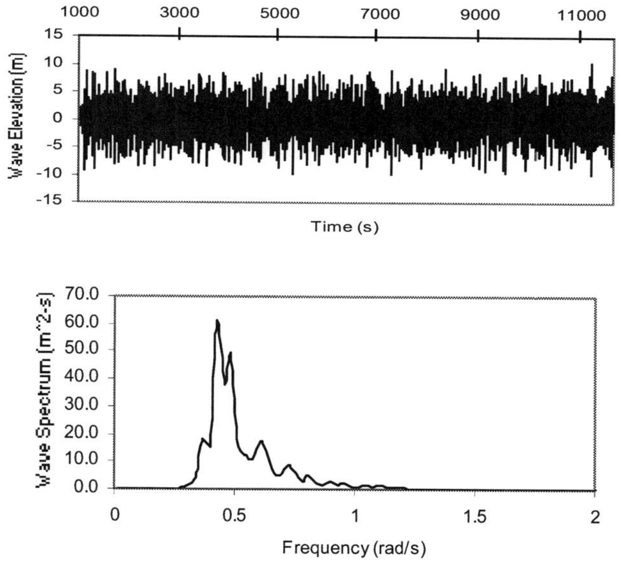

The time series and spectrum of wave elevation

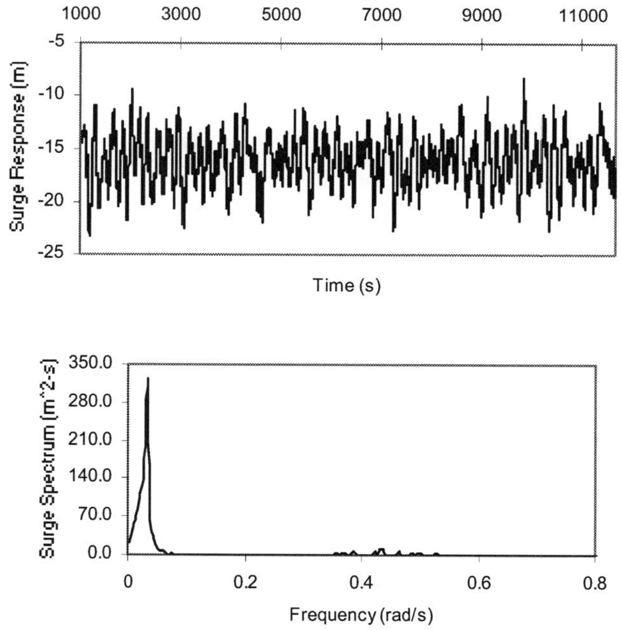

The time series and spectrum of surge responses

Fig. 3-8 The time histories and spectra of wave elevation, surge, heave, pitch motions, and tension of leg 2 (continued to page 3.19)

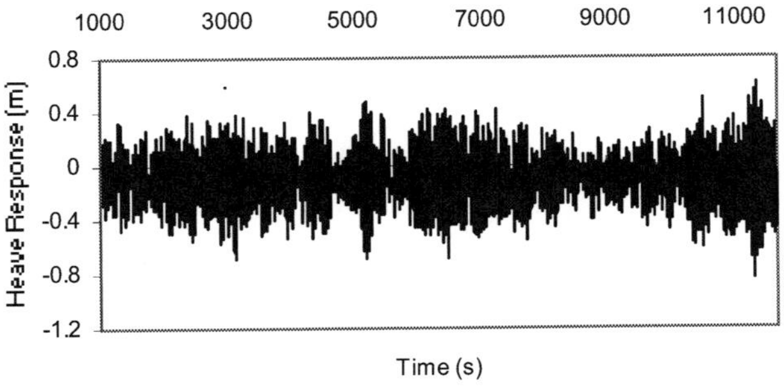

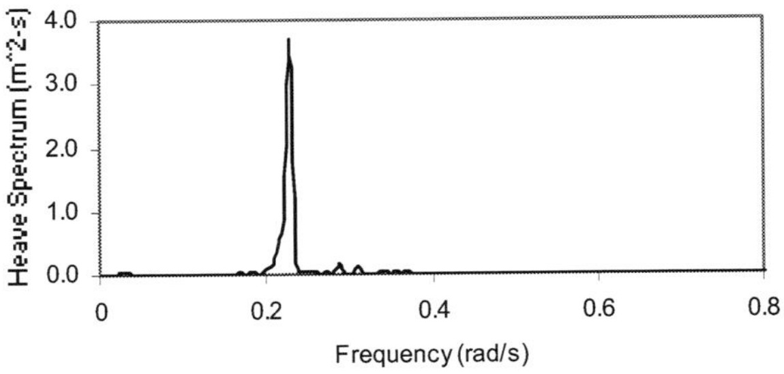

The time series and spectrum of heave responses

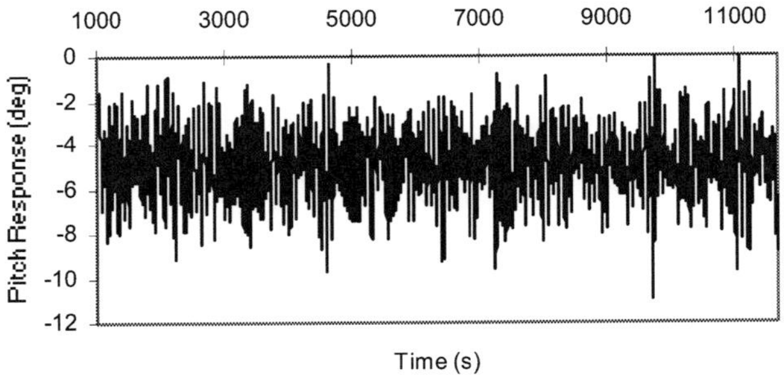

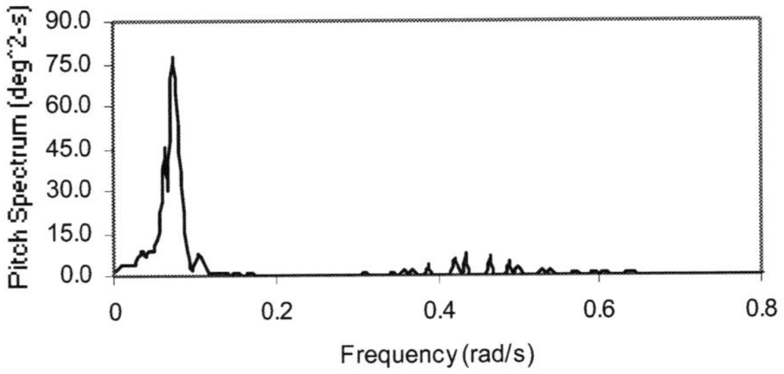

The time series and spectrum of pitch responses

Fig. 3-8. The time histories and spectra of wave elevation, surge, heave, pitch motions, and tension of leg 2 (continued)

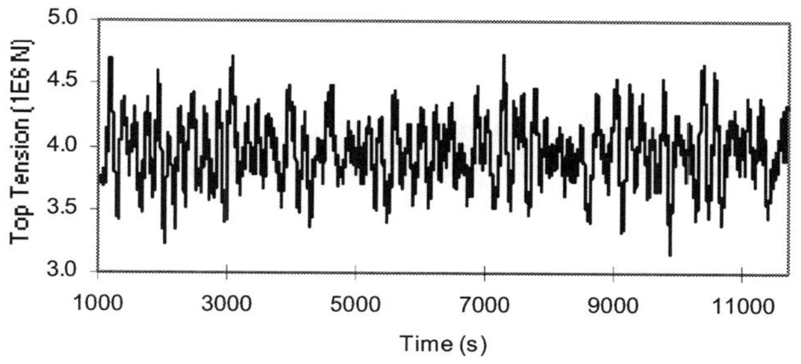

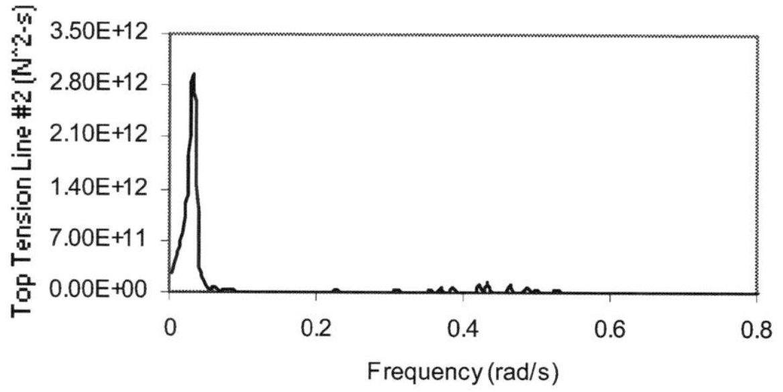

The time series and spectrum of tension for leg 2

Fig. 3-8. The time histories and spectra of wave elevation, surge, heave, pitch motions, and tension of leg 2 (continued)

Table 3-6. 6000 ft.-Spar Responses
(Hurricane case, standard=standard deviation)

Motion	Mean	Wave-freq. standard	Low-freq. standard	Total standard	max
Surge	-16.2	0.6	2.2	2.3	-23.4
Heave	-0.09	0.2	0.06	0.21	-0.84
Pitch	-4.9	0.5	1.4	1.5	-11
Yaw	0.0	0.0	0.04	0.04	-0.19

The surge mean offset can be roughly predicted from the total surge mean force divided by surge stiffness. The resulting surge offsets generally coincide with the simple estimation. It is interesting to note that the low-frequency standard deviation is greater than that of wave frequency in all modes except heave, which can also be confirmed from Figure 3-6. In the pitch calculation shown in Table 3-5, the buoyancy can effects are not included, and thus the maximum pitch is appreciably over predicted.

Table 3-7. 6000 ft.-Spar Mooring Top Tension
(Hurricane case)

Leg	Mean	Wave-freq. standard	Low-freq. standard	Total standard	Max
#1	39.5	0.7	2.2	2.3	47.1
#7	14.1	0.5	0.9	1.0	18.3
#2	39.6	0.6	2.4	2.5	47.3

The maximum static and dynamic tension occurs at the up-wave mooring, either mooring #1 or mooring #2. The ratio of maximum tension to breaking strength is 37 percent. The ratio of maximum tension to static tension is about 1.2, and most of the dynamic part comes from the slowly varying component. Therefore, the taut polyester lines behave in a quasi-static manner. This trend may not be true for semi-taut steel lines. The contribution of wave-frequency tension is greater in slack lines.

Sensitivity Study and Discussions

The following cases are selected for sensitivity analysis. The hull drag coefficient used in all the simulations is 1.15 except for Case B.

Case A: Hull/mooring/riser fully coupled analysis with the prescribed model (drag coefficient of spar hull = 1.15)

Case B: Hull/mooring/riser fully coupled analysis with the prescribed model (drag coefficient of spar hull = 0.6)

Case C: Uncoupled quasi-static analysis; all mooring lines and risers are massless but have nonlinear stiffness as in Figure 3-4; that is, there is no dynamic effect by mooring-lines or risers. The mooring tension is determined by surge-tension relation like Figure 3-4.

Case D: Hull/mooring/riser fully coupled analysis without wind; by comparing Cases A and D, the wind effects can be explicitly seen.

Case E: Hull/mooring/riser fully coupled analysis without current; by comparing Cases A and E, the current effects can be explicitly seen.

Case F: All the 23 risers are replaced by a single equivalent massless riser with the same total tension 28,325 kips (12.6×10^7 N). The resulting surge/sway stiffness at the keel is approximately calculated from tension/length. The equivalent riser stiffness matrix is added to the hydrostatic-stiffness matrix.

Case G: Semi-coupled dynamic analysis; model mooring and risers as massless, nonlinear spring, calculate hull responses, input the calculated motions at the fairlead, and run line dynamics program for each mooring line. Compared to C, the dynamic effects of each mooring line can be seen.

Case H: Hull/mooring/riser fully coupled analysis with equivalent static wind loading based on 1-min. mean wind velocity. By comparing A and H, the effect of dynamic wind can be deduced.

Case I: Hull/mooring/riser fully coupled analysis (same as Case A) with additional simplified buoyancy-can modeling (riser model 2) in the roll/pitch hydro-static coefficients. By comparing I and A, the effect of air cans as inverted pendulum can be observed.

Case J: Hull/mooring/riser fully coupled analysis (same as Case A) with riser model 3. Risers are extended through riser guides inside moonpool and modeled by FEM elements. Buoyancy cans are attached at the top.

The numerical results for all the cases are summarized in Table 3-7. First of all, we found that the wave-frequency standard values are not sensitive to the variation from one case to another, as can be seen in Table 3-4. This fact was also confirmed by Colby et al. (2000) and Ma et al. (2000). Therefore, the ensuing discussion will primarily be focused on the slowly-varying components.

- ***The sensitivity against hull drag coefficient***

The hull and mooring/riser drag forces are difficult to model accurately with the small-scale model testing due to the mismatch of Reynolds numbers between the model and prototype. This distorted scale problem may be more serious when the mass of the slender members becomes increasingly appreciable compared to the platform itself as water depth increases. It is well known that the viscous effects are usually exaggerated in the model testing; thus, the small-scale experimental results are in general less conservative. To check the effect of viscous drag forces, we reduced the drag coefficient on the hull almost to half in Case B and ran the same coupled-analysis code. In this test, the drag coefficients of mooring/riser were not varied. When Case B is compared against Case A, the mean surge offset is reduced by 14 percent, while the standard surge is increased by 12 percent—the combined effect of which results in 9 percent decrease in maximum surge and only 1 percent decrease in maximum mooring tension. The increase of the standard surge is due to the decrease in viscous damping on the hull.

- ***The sensitivity against uncoupled quasi-static or semi-coupled dynamic approach***

Offshore industry frequently uses simplified, uncoupled quasi-static approach in mooring design. In a typical quasi-static analysis, the surge response is first calculated by modeling the mooring lines by massless linear or nonlinear springs; then the mooring line tension is estimated from the surge static-offset curve. In general, additional coupling contributions from damping as well as current loading on the mooring/riser need to be assessed and given as input to the uncoupled quasi-static analysis to improve the final result. However, no such contributions are added here. Kim et al. (1999) showed that in the case of a truss spar at 3240-ft. water depth, the quasi-static approach can significantly underestimate the maximum mooring-line tension. It is particularly so when mooring lines are semi-taut and their natural frequencies are close to the incident wave frequencies.

To test the reliability of the uncoupled quasi-static approach in the present problem, all mooring lines and risers are modeled by massless nonlinear spring, as in Figure 3-4; that is, all the inertia forces and hydrodynamic forces on the mooring/riser are neglected. After calculating the surge time series, the corresponding mooring tension time history can straightforwardly be constructed from Figure 3-4.

At this stage, if the hull motions are input at the fairlead and the dynamics program is run for each mooring line; then it is called a semi-coupled dynamic approach. When the dynamic effects of mooring lines are significant, the semi-coupled dynamic approach should be more reliable than the uncoupled quasi-static approach. In this example, however, slowly varying surge responses are dominant over wave-frequency components; therefore, it is expected that the mooring tension is likely to behave in a quasi-static manner.

Consequently, Case C (uncoupled quasi-static approach) is almost identical to Case G (semi-coupled dynamic approach). The 21 percent increase of standard surge in Case C and G compared to Case A should be attributed to the absence of viscous damping from mooring lines and risers. The reduction in the mean surge offset is only 1 percent. The standard deviation in pitch is also increased by 28 percent in Case C and G by the same reason. The observed trend is generally similar to that of Colby et al. (2000). The maximum mooring tension in Case C and G is about 5 percent larger than Case A mainly due to the increase in surge. Even though line dynamics are ignored, this example shows that the mooring tension can be overestimated by an uncoupled quasi-static approach by ignoring riser/mooring damping. This rather surprising result may happen particularly when line dynamic effects are small. It is clear that the viscous damping from mooring lines and risers plays an important role in the dynamics of deepwater platforms.

- ***The sensitivity against wind***

The comparison of Case A (fully coupled analysis with dynamic wind) and Case D (fully coupled analysis without wind) clearly shows the effect of wind loading. First, the mean surge offset is reduced by 67 percent and the standard surge by 32 percent. The surge standard deviation is reduced mainly due to the reduction of the low-frequency part. In other words, the wind loading is very important to the mean and slowly varying surge responses. The wind also significantly influences pitch motions because its center of pressure is located far from the center of rotation. For instance, the mean pitch angle is reduced from 4.9°to 1.1° and the maximum pitch from 11° to 5.7° when the wind is absent. The removal of wind loading also results in a 32 percent reduction in the maximum tension on the highest-tension member.

When the dynamic-wind-generation scheme is not available in the model testing, constant string forces are frequently applied at the center of pressure to at least model the static part in a reasonable manner. Here, 1-minute mean wind velocity is used. The results of this static-wind modeling are given in Case H where the mean offset is close (8 percent larger) to the actual value but the surge standard value is significantly underestimated (29 percent smaller) like Case D without wind. Similar phenomenon also can be observed for the mean and standard pitch responses. When 1-hour mean wind velocity is used instead, the error for the mean surge offset is increased to 15 percent.

- ***The sensitivity against current***

The comparison of Case A (fully coupled analysis with current) and Case E (fully coupled analysis without current) clearly shows the effect of current on spar dynamics. The mean surge offset is decreased by 29 percent, while the total standard deviation in surge is increased by 15 percent. The reduction of the mean offset is due to the absence of current-induced viscous drag force on the hull and mooring/riser. On the other hand, the increase of standard surge is due to the decrease in viscous damping in the absence of current.

When the loop-current environment is considered, the effects of currents play a much more important role. In particular, the hull and line VIVs are additionally generated, which makes the motion characteristics more complicated. The typical hull VIV patterns include figure 8-shaped and banana-shaped motions in the lateral direction perpendicular to the current direction. These VIVs can significantly amplify the line dynamic tension. The VIV can appreciably be reduced by attaching continuous or discontinuous spiral strakes on the hull. It is also well known that the VIVs may significantly (up to the factor of two) increase the in-line drag forces. Therefore, VIVs are generally important in spar global motion analyses in the presence of strong currents. However, it is still very difficult to predict the VIVs both in numerical analysis and model-scale experiment. Numerically, the 3D Navier-Stokes equation has to be solved at a very large Reynolds number with sophisticated turbulence modeling. Experimentally, there always exist scale effects, and the VIV patters vary in complicated manners depending on size, flow conditions, and strake arrangement.

- ***The sensitivity against equivalent linear riser modeling***

In Case F, the entire riser group is replaced by the riser-stiffness matrix, which is added to the hydrostatic-coefficient matrix, as described in the preceding section. Compared to Case A, the mean surge offset is slightly decreased due to zero current force on risers. The amount of decrease is very small because the top portion of risers is protected by hollow cylindrical hull and not directly exposed to the hurricane-induced currents, which decay relatively rapidly with depth. On the other hand, the standard surge response was increased by 15 percent due to smaller viscous damping in the absence of risers. In particular, the wave-frequency surge motion nearly remains the same, while the low-frequency surge response is appreciably increased. This occurs because the slowly varying surge response is due to the resonance and, thus, is sensitive to the available amount of damping. We can also notice that the pitch standard value is increased by 27 percent by using equivalent massless riser modeling.

- ***The sensitivity against air cans effect***

The comparison of Case A (fully coupled analysis with risers model 1) and Case I (fully coupled analysis with riser model 2) clearly shows the effect of additional restoring moment by air cans in pitch and roll modes. As mentioned earlier, the portion of risers above the keel was first modeled in riser model 2 as a simple inverted pendulum, which contributes additional pitch and roll stiffness. As a result, the mean pitch offset is reduced by 59 percent and the rms surge by 32 percent. The rms-pitch reduction is mainly due to the increase of the roll/pitch natural frequency, which makes the system stiffer and reduces the low frequency responses. In particular, the shift of natural frequency makes the wind loading less influential to the slowly varying pitch responses. Due to the reduction both in mean and rms values, the maximum pitch in 3 hours simulation is reduced by 47 percent.

In Case J (riser model 3), risers are extended inside the moonpool, and more rigorous FEM modeling is developed for that portion. The results of Case J are almost identical to those of Case I, which implies that the simple approach proposed here for air-can effects works very well.

It is also reported by Koo (2003) that the gap contact between buoyancy cans and riser guides can appreciably reduce the heave motion through Coulomb friction.

Table 3-6. Comparison of Statistic for Cases A-J

Case	Code	Surge Total			Surge low-freq	Surge wave-freq
		Mean	St.dev	Max	St.dev	St.dev
Fully Coupled (Cd = 1.15)	A	-16.18	2.33	-23.36	2.24	0.65
Fully Coupled (Cd = 0.6)	B	-13.98	2.61	-21.28	2.52	0.65
Quasi-static (with moor & riser)	C	-15.99	2.81	-24.28	2.74	0.65
Fully Coupled (without wind)	D	-5.39	1.58	-10.99	1.44	0.65
Fully Coupled (without current)	E	-11.54	2.69	-19.45	2.61	0.65
Coupled (with equivalent riser, Cd=1.15)	F	-15.54	2.58	-22.82	2.50	0.65
Semi Coupled	G	-15.99	2.81	-24.29	2.74	0.65
Fully Coupled (static wind – 1 minute)	H	-17.47	1.66	-23.60	1.53	0.65
Fully Coupled with risers model 2	I	-16.52	2.36	-23.86	2.27	0.65
Fully Coupled with risers model 3	J	-16.58	2.34	-23.93	2.25	0.64

Case	Code	Heave			Heave low-freq	Heave wave-freq
		Mean	St.dev	Max	St.dev	St.dev
Fully Coupled (Cd = 1.15)	A	-0.09	0.21	-0.84	0.06	0.20
Fully Coupled (Cd = 0.6)	B	-0.07	0.21	-0.82	0.06	0.20
Quasi-static (with moor & riser)	C	-0.10	0.24	-0.89	0.06	0.23
Fully Coupled (without wind)	D	0.00	0.22	-0.79	0.05	0.21
Fully Coupled (without current)	E	-0.06	0.21	-0.80	0.06	0.20
Coupled (with equivalent riser, Cd=1.15)	F	-0.09	0.21	-0.82	0.06	0.20
Semi Coupled	G	-0.01	0.24	-0.88	0.07	0.23
Fully Coupled (static wind – 1 minute)	H	-0.11	0.20	-0.86	0.06	0.19
Fully Coupled with risers model 2	I	-0.09	0.21	-0.85	0.06	0.20
Fully Coupled with risers model 3	J	-0.01	0.21	-0.85	0.06	0.20

Case	Code	Pitch Total			Pitch low-freq	Pitch wave-freq
		Mean	St.dev	Max	St.dev	St.dev
Fully Coupled (Cd = 1.15)	A	-4.91	1.48	-10.96	1.38	0.55
Fully Coupled (Cd = 0.6)	B	-4.51	1.65	-11.25	1.56	0.55
Quasi-static (with moor & riser)	C	-4.99	1.89	-12.49	1.81	0.55
Fully Coupled (without wind)	D	-1.09	0.95	-5.74	0.78	0.54
Fully Coupled (without current)	E	-4.22	1.70	-10.87	1.61	0.55
Coupled (with equivalent riser, Cd=1.15)	F	-5.05	1.88	-12.37	1.80	0.55
Semi Coupled	G	-4.99	1.89	-12.49	1.81	0.55
Fully Coupled (static wind – 1 minute)	H	-5.31	0.95	-10.06	0.78	0.54
Fully Coupled with risers model 2	I	-2.00	1.01	-5.85	0.84	0.57
Fully Coupled with risers model 3	J	-2.31	1.08	-6.22	0.92	0.56

Case	Code	Highest Tension			Higher Tension low-freq	Highest Tension wave-freq
		Mean	St.dev	Max	St.dev	St.dev
Fully Coupled (Cd = 1.15)	A	3.97E+06	2.47E+05	4.73E+06	2.39E+05	6.42E+04
Fully Coupled (Cd = 0.6)	B	3.81E+06	2.78E+05	4.67E+06	2.70E+05	6.40E+04
Quasi-static (with moor & riser)	C	3.95E+06	3.02E+05	4.96E+06	2.93E+05	6.93E+04
Fully Coupled (without wind)	D	2.74E+06	1.35E+05	3.22E+06	1.20E+05	6.23E+04
Fully Coupled (without current)	E	3.64E+06	2.83E+05	4.49E+06	2.76E+05	6.37E+04
Coupled (with equivalent riser, Cd=1.15)	F	3.91E+06	2.76E+05	4.77E+06	2.68E+05	6.42E+04
Semi Coupled	G	3.95E+06	3.00E+05	4.93E+06	2.93E+05	6.50E+04
Fully Coupled (static wind – 1 minute)	H	4.11E+06	1.51E+05	4.67E+06	1.37E+05	6.41E+04
Fully Coupled with risers model 2	I	3.94E+06	2.47E+05	4.69E+06	2.38E+05	6.46E+04
Fully Coupled with risers model 3	J	3.95E+06	2.45E+05	4.70E+06	2.36E+05	6.44E+04

Survivability Issues

During 2004-05, three category 5 hurricanes—Ivan, Katrina, and Rita—hit numerous floating platforms in the Gulf of Mexico. Surprisingly, most of the floating production platforms and mooring system survived, while major damage occurred on super-structures above deck. For example, an upper-derrick and its substructure on Medusa truss spar were shifted and toppled during Hurricane Ivan.

With the time-domain simulation program, the 6DOF acceleration time series at the derrick and the corresponding instantaneous inertial loading can be simulated. With the wind-velocity time series, the instantaneous wind loading on derrick structures can also be obtained. In addition, the instantaneous rotational-motion-induced gravitational/centrifugal forces can be added. They can then be used directly to simulate the slide/shear forces and moments at the derrick and substructure footing to check their survivability in harsh sea condition. In case of spars, the acceleration related to rotational motions can be significantly increased as the sea state approaches the 1000-year hurricane condition (Ward et al. 2007). For instance, the acceleration at a typical spar derrick during the hurricane Ivan can be as large as 0.5 g, while those for 10-year and 100-year hurricanes are 0.2 g and 0.4 g, respectively.

For the 1000-year storm, the peak wavelength is much bigger than that of the 100-year storm, and the heave motion of spar is expected to significantly increase due to the deeper penetration of particle movement by longer waves. The increased heave motion will directly influence the wave-frequency dynamic loading on mooring lines. As a result, the dynamic tension of a spar in 1000-year condition can be significantly higher than that of the 100-year storm. However, the effect may be compensated by the increase of Coulomb friction between riser guide and buoyancy cans as well as increased nonlinear viscous and mooring damping, which can all be included in the present nonlinear time-domain global motion analysis.

With the present time-domain program, the progressive failure of mooring lines can also be simulated. When one line fails, there may be strong transient over-shooting in hull motions that may further trigger the breakage of other lines. It is also well known that the suction-pile anchor is not strong against out-of-plane loading. When this happens, the mooring line will fail not at the fairlead but at the anchor. When a mooring line fails at the anchor, the platform may eventually drift with the mooring line hanging on the hull and dragging on the seafloor. These kinds of phenomena can be analyzed straight forwardly by the hull-mooring-riser coupled time-domain analysis program, as explained in this paper.

Summary and Conclusions

Nonlinear hull/mooring/riser coupled dynamic analyses of a classic spar designed for 6000-ft. water depth are conducted in the time domain for the 100-year hurricane condition with non-parallel wind, wave, and current. A comprehensive sensitivity study against various analysis/environment parameters was carried out to better understand the underlying physics and the role of each parameter.

In all cases considered, the wave frequency components remain almost the same, while the slowly varying motions change case by case. It is found that the mooring lines and risers contribute appreciably in surge/sway damping. In this example the standard values of fully coupled analysis are smaller than those of uncoupled quasi-static analysis and semi-coupled dynamic analysis, which can be attributed to the additional mooring-line damping in fully coupled analysis. The quasi-static analysis looks reasonable in the studied case since the slowly varying motions are dominant over wave-frequency motions and the mooring lines behave in a quasi-static manner. The larger hull drag coefficient resulted in larger mean offsets but smaller dynamic responses, as expected. It is particularly underscored that the equivalent static wind modeling may lead to significant underestimation of slowly varying surge/sway and pitch/roll responses. The total removal of wind loading resulted in a 32 percent reduction in the maximum mooring tension. The removal of currents resulted in a 32 percent reduction in mean surge and 15 percent increase in standard surge.

The additional roll- and pitch-restoring moments by air-can resulted in smaller mean offsets and dynamic responses. As a result, the maximum pitch of the case with buoyancy cans is reduced by about a factor of two compared to that of the case without buoyancy cans. This fact was confirmed both in simple and more rigorous riser models inside moonpool. It is particularly underscored that the riser model without proper air-can effect may lead to significant overestimation of mean and maximum pitch/roll responses. The gap contact forces between buoyancy cans and riser guides also cause Coulomb friction in the vertical direction, which in turn appreciably reduces the spar heave motion, as illustrated in Koo (2003).

References

Arcandra (2001). "Hull/mooring/riser coupled dynamic analysis of a deepwater floating platform with polyester lines." *Ph.D. Thesis*, Texas A&M University.

Cao, P. & Zhang, J. (1996). "Slow motion responses of compliant offshore structures." *Proc. 6th ISOPE'96 Conf.,* Los Angeles, 296-302

Celebi, M.S., Kim, M.H., & Beck, R.F. (1998)."Fully nonlinear 3D numerical wave tank simulations." *Journal of Ship Research,* Vol.42, No.1, 33-45

Clark, P.J., Malenica, S. & Molin, B. (1993). "An heuristic approach to wave drift damping." *Journal of Applied Ocean Research*, Vol.15, 53-55

Colby, C., Sodahl, N., Katla, E., Okkenhaug, S. (2000). "Coupling effects for a deepwater spar." *OTC'2000* #12083, Houston.

Eatock Taylor, R., Rainey, R.C.T. & Dai, D.N. (1992). "Nonlinear hydrodynamic analysis of TLPs in extreme waves, slender body and diffraction theories compared." *Proc. Behavior of Offshore Structures*, London

Emmerhoff, O.J. & Sclavounos, P.D. (1992). "The slow drift motion of arrays of vertical cylinders." *Journal of Fluid Mechanics*, Vol.242, 31-50

Faltinsen, O.M. (1990). "Sea loads on ships and offshore structures." *Cambridge University Press*

Faltinsen, O.M. (1994). "Wave and current induced motions of floating production systems." *Journal of Applied Ocean Research*, Vol.15, 351-370

Garrett, D.L. (1982)."Dynamic analysis of slender rods." *J. Energy Resources Tech. ASME Trans*. Vol.104 302-307

Goupta H, Finn, L, Weaver, T. (2000). "Effects of Spar Coupled Analysis." *OTC'2000* # 12082 2000

Halkyard, J.E. (1996). "Status of spar platforms for deepwater production systems." *Proc. of ISOPE'96*, Los Angeles.

Kim, C.H. (1995). "Recent progress in numerical wave tank research: a review." *Proc. 5th International Offshore & Polar Engineering Conf.,* Vol.3, 1-9, The Hague

Kim, D.J. & Kim, M.H. (1995). "Interaction of a large 3D body with waves and currents by THOBEM." *Proc. 10th Workshop on Water Waves & Floating Bodies*, Oxford

Kim, M.H., Ran, Z., & Zheng, W. (1999). "Hull/mooring coupled dynamic analysis of a truss spar in time domain." *Proc. of ISOPE'99*, Brest, France.

Kim, M.H. & Yue, D.K.P. (1989a). "Slowly-varying wave drift forces in short-crested irregular seas." *Journal of Applied Ocean Research*, Vol.11, 2-18

Kim, M.H. & Yue, D.K.P. (1991). "Sum- and difference-frequency wave loads on a body in uni-directional Gaussian seas." *Journal of Ship Research*, Vol.35, 127-140

Kim, M.H. (1992). "Difference-frequency wave loads on a large body in multi-directional waves." *Journal of Applied Ocean Research*, Vol.14, 353-370

Kim, M.H. & Chen, W. (1994)."Slender body approximation for slowly-varying wave loads in multi-directional waves." *Journal of Applied Ocean Research*, Vol.16; 141-163

Kim, Y.B. (2003)."Dynamic analysis of multiple body floating platforms coupled with mooring lines and risers." *Ph.D. Thesis*, Texas A&M University.

Koo, B.J. (2003)."Evaluation of the effect of contact between risers and guide frames on offshore spar platform motions." *Ph.D. Thesis*, Texas A&M University.

Koo, B.J., Kim, M.H., and Randall, R. (2004)."The effects of nonlinear multi-contact coupling with gap between risers and guide frame on global spar motion analysis." *Journal of Ocean Engineering*, Vol.31, 1469-1502

Korsmeyer, F.T., Lee, C.H., Newman, J.N. & Sclavounos, P.D. (1988). "The analysis of wave effects on tension-leg platforms." *Proc. OMAE*, Houston

Lee, C.H., Newman, J.N., Kim, M.H. & Yue, D.K.P. (1991). "The computation of second-order wave loads." *Proc. of OMAE'91*, Stavanger, Norway.

Ma, W., Lee, M.Y., Zou, J., and Huang, E.W. (2000). "Deepwater nonlinear coupled analysis tool." *OTC'2000* #12085, Houston.

Mekha, B.B., Weggel, D.C., Johnson, C.P. & Roesset, J.M. (1996). "Effects of second-order diffraction forces on the global response of spars." *Proc. of 6th ISOPE Conf.,* Los Angeles, 273-280

Newman, J.N. (1974). "Second-order slowly varying forces on vessels in irregular waves." *Symp. on Dynamics of Marine Vehicles and Structures in Waves*, London.

Pauling, J.R. & Webster, W.C. (1986). "A consistent large amplitude analysis of the coupled response of a TLP and tendon system." *Proc. 5th OMAE Conf.*, Tokyo, Vol.3, 126-133

Rainey, R.C.T. (1989). "A new equation for calculating wave loads on offshore structures." *Journal of Fluid Mechanics*, Vol.204, 295-324

Ran, Z. (2000). "Coupled dynamic analysis of floating structures in waves and currents." *Ph.D. Thesis*, Texas A&M University.

Ran, Z., Kim, M.H., Niedzwecki, J.M., & Johnson, R.P. (1996). "Response of a spar platform in random waves and currents (experiment vs. theory)." *Journal of Offshore & Polar Engineering,* Vol.6. No.1, 27-34

Ran, Z. & Kim, M.H. (1996). "Nonlinear coupled response of a tethered spar in waves." *Proc. International Offshore & Polar Engineering Conf.,* Los Angeles

Steen, Irani, M., & Kim, M.H. (2004)."Prediction of spar responses, model test vs analysis." *Proc. 2004 OTC,* Houston

Ward, E.G., Gebara, J.M., Kim, M.H., and Ghoneim, N. (2007)."Performance of drilling rig sea fastenings on floating production systems." *Proc. Offshore Technology Conference,* paper # 18986

Webster, W.C. (1995). "Mooring induced damping." *Journal of Ocean Engineering,* Vol.22, No.6, 571-591

Wichers, J.E.W. & Devlin, P.V. (2001)."Effect of coupling of mooring lines and risers on the design values for a turret-moored FPSO in deep water of GOM." *Proc. Of 11th ISOPE,* Vol.3, 480-487

Chapter 4: Spar Hull/Mooring/Riser Coupled Dynamic Analysis, VIM Effects, and Mathieu Instability

By Jun Zou, Ph.D., Manager of Naval Architecture, Houston Offshore Engineering

Abstract

This article presents a study of hull/mooring/riser coupled dynamic analysis for the conventional spar. Spars are a popular floating production system (FPS) with a very deep draft. The mooring system is conventional chain-wire-chain mooring, and the top tensioned riser (TTR) is air can supported. Coupled dynamic analyses are becoming increasingly important in deep and ultra deep water field developments because (1) more pronounced interactions of spar hull with its moorings and risers have been observed and (2) the limitation of the existing test facilities to carry out physical model tests without mooring/riser truncation with proper model scale has been recognized. One of the objectives of this article is to identify the characteristics of the spar dynamic responses in extreme conditions.

The conventional spar hull is cylindrical in cross-section with a central moonpool for the riser system. The spar riser system typically consists of the risers themselves, tensioning system, supporting guide frame inside the moonpool, and keel joints. Spar hull/mooring/riser coupled analysis includes the effects between hull/moorings, risers/supporting guide frames, and keel joints inside the moonpool and hull/risers outside the spar hull. The truly coupled analysis as stated in this article means mass, damping, stiffness, and load coupling from the hull/mooring/riser. For purposes of detailed design, it is necessary to model the contact of risers and supporting guide frames with realistic boundary conditions. The spar hull/mooring/riser coupled analysis results and physical model test measured results are presented and compared.

The conventional spar hull is a cylindrical structure with the larger length over diameter ratio. It has been reported that serious vortex induced motions (VIM) have been observed in field measurements. On the basis of the model test results, VIM effects on spar hull motions, mooring line tensions, riser tensions, and strokes are highlighted.

Because of spar pitch/heave coupling, Mathieu instability may become excited if certain criteria can be satisfied. The instability may disappear with the presence of damping due to hull/mooring/riser coupled effects. Mathieu instability could be triggered in the light damped system, which does not necessarily involve regular waves. It is worth checking Mathieu instability in the design to exclude any possibility of instability problems. One of the objectives of this article is to investigate damping effects on suppressing Mathieu instability.

Introduction

A conventional spar hull consists of a hard tank, a skirt tank, and a keel tank with a draft of 650 ft. The hard tank provides buoyancy, the skirt tank is usually flooded with water to entrap a large amount of mass, and the keel tank is filled with solid ballast and entrapped water. The spar hull is cylindrical with a central moonpool for the riser system. The spar riser system typically consists of the risers themselves, the tensioning system, and a supporting guide frame inside the moonpool. The guide frames and keel joints are used to constrain the riser motions in the hull's transverse direction. They also allow the hull to move relative to the risers in the hull's longitudinal direction, including friction forces caused by contact of the supports and riser system. An alternative tensioning system for spar top tensioned risers was proposed by Finn et al. (2001). This concept introduces a significantly higher coupling in heave, and as a result, heave restoring as well as the heave natural period will be influenced by the riser system. In this study, only air can supported riser systems have been studied and modeled. When the spar oscillates around the mean offset position, the buoyancy of air cans provides additional restoring moment to resist the pitch/roll motions of the hull. Contact forces at the keel joints and other riser support frames induce friction to resist hull heave motions. The contact forces at different levels along the spar hull longitudinal generate resistance moments to affect pitch/roll motions. These lateral contact forces affect surge/sway motions of the hull in a similar manner.

Spar coupled time domain analysis technologies (Kim et al. 1997 and 2001; Ma et al. 2000; Gupta et al. 2000; Zhang and Zou 2002) have been reported. However, Kim et al. (1997 and 2001) and Ma et al. (2000) ignored the coupling effects of risers/supporting guide frames and keel joints inside the spar moonpool. Only Gupta et al. (2000) and Zhang and Zou (2002) considered these effects. Gupta et al. (2000) used two commercial programs to model the hydrodynamics and risers/moorings separately and then combined them in the third program. Thus, many data communications are required between the two different commercial programs. This additional step slows down the numerical simulation and may induce potential errors, which would be difficult to detect. Zhang and Zou (2002) have developed an integrated program to perform spar hull/mooring/riser coupled dynamic analysis effectively. These developments and validations are outlined in this article.

A deepwater nonlinear coupled analysis tool (DeepCAT), jointly developed by ABS and ABB in 1999, has the ability to simulate platform motions and associated mooring/tendon and risers, including second-order effects due to wind, current, and waves. Coupled analysis techniques, modeling methods of large volume and slender member structures, environmental conditions, and examples of application were summarized in Ma et al (2000). On the basis of the previous work, new features have been added to consider the effects of contact forces of risers and supporting guide frames and keel joints with realistic boundary conditions.

The spar concept is an FPS with a very deep draft. It is well known that the pitch motion of a spar coupled with heave can be recast into a standard damped Mathieu's equation. Haslum and Faltinsen (1999) showed a stability diagram for Mathieu's equation without considering pitch damping effects. Zhang et al. (2002) extended their studies to include damping effects and develop a new stability diagram for damped Mathieu's equation. With a newly developed stability diagram, the instability problems of a classical spar in the principal and secondary unstable zones were examined. One of the goals of this article is to explore the role of damping due to hull/mooring/riser coupling effects on suppressing Mathieu instability.

In this article, the spar hull/mooring/riser coupled time domain dynamic analysis techniques are reviewed. Then damped Mathieu instability evaluation methods are outlined. Descriptions of the study case; applications of the newly developed analysis techniques; VIM effects on spar hull motions, mooring line tensions, riser tensions, and strokes; a damped stability diagram to the spar; and analytical results and validations follow. Finally, conclusions are drawn and recommendations are given.

Spar Hull/Mooring/Riser Coupled Dynamic Analysis

The coupled dynamic analysis technique (hull/mooring and hull/riser outside of the hull) has been presented in Ma et al. (2000). This analytical capability has been further extended to include contact forces of risers/supporting guide frames and keel joints and has been demonstrated in Zhang and Zou (2002). Additionally, spar hull/mooring/riser coupled dynamic analysis techniques are reviewed.

Coupled Dynamic Equation

A 6DOF coupled dynamic equation of motion is used as follows

$$[M]\{\ddot{U}\}+[C]\{\dot{U}\}+[K]\{\Delta U\}=\{F\}+\{F_m\} \tag{4-1}$$

where

[M] = mass and inertia matrix (6 × 6), structure mass and inertia + added mass and inertia of the platform,

[C] = damping matrix (6 × 6), the potential damping + viscous damping + wave drift damping on the platform,

[K] = stiffness matrix (6 × 6), hydrostatic stiffness (heave and roll/pitch) + stiffness due to moorings and risers,

{F} = load vector (6 × 1) of first- & second-order wave loads + viscous loads + wind dynamic loads, and/or other applied loads,

$\{F_m\}$ = load vector (6 × 1) of mooring/tendon, and riser tensions at the connected locations,

{U } = unknown motion vector (6 × 1) in the sequence of surge, sway, heave, roll, pitch, and yaw, respectively.

Coupled Dynamic Analysis Flow Chart

A spar hull/mooring/riser coupled dynamic analysis flow chart is shown in Figure 4-1. The first- and second-order diffraction and radiation problems are solved in the frequency domain and then are transferred into the time domain by applying two terms of the Volterra functional polynomial model. Random wind dynamic forces are generated by an API wind spectrum. Line (mooring/tendon, riser) dynamic tensions are computed by solving a set of algebraic equations and are transferred to the platform. Figure 4-1 shows the exchange platform motions and line dynamic tensions. After assembling the mass matrix, damping matrix, and stiffness matrix, and the exciting load vector, the governing Eq. 1 can be solved to obtain motions (surge, sway, heave, roll, pitch and yaw) at the center of gravity (CG) of the platform.

Line Dynamics

The line dynamics should be interpreted either as mooring line dynamics or riser dynamics. The formulation of the dynamics of inextensible slender rods was first introduced by Garrett (1982) and then expanded by Paulling and Webster (1986) to consider stretch and various loads to the real dynamic problem. The detailed line dynamic formulations can be found in Appendix I. After discretizing the vector governing equations into algebraic equations by the Galerkin method, the original vectored equations are solved using the second-order Adams-Moulton integration algorithm (Ma and Webster 1994).

Sub-System or Component Modeling

Spar Hull Modeling

The spar hull is modeled using a panel-based diffraction theory including first- and complete second-order wave exciting loads, for example, WAMIT. Viscous loads on spar hull were accounted for by Morison's equation.

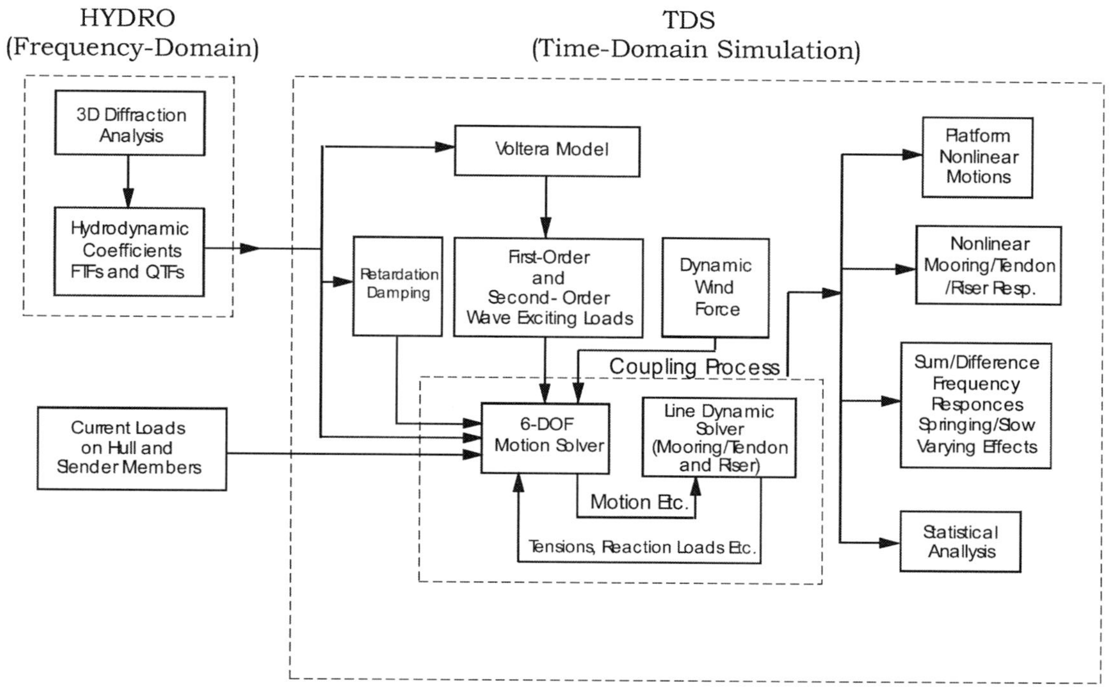

Fig. 4-1. Coupled dynamic analysis flow chart

Mooring Line and Riser Modeling

Slender members, such as mooring lines and risers, are modeled by beam elements. Viscous and inertia loads on these slender members are accounted for by Morison's equation. Their line dynamics of these slender members are simulated in the coupled dynamic analysis. The detailed numerical formulations can be found in Appendix I.

Mooring Line Coupling Modeling

As shown in Figure 4-1, mooring line tensions and hull motions are exchanged back and forth between the spar hull and its mooring lines. For boundary conditions between the hull and mooring line, an ideal hinge is assumed.

Riser Air Cans and Their Supporting Guide Frame Modeling

Some assumptions for riser air cans and their supporting guide frame modeling are summarized as follows:

- There is point contact between the air can and its supporting guide frame at the contact position due to the round shape of the supporting guide frame. One air can may contact a few supporting guide frames at different elevations. Thus, the supporting guide frames can be modeled by a set of contact points that move rigidly with the spar hull.
- The node in which the air can comes into contact with its supporting guide frame moves with the spar hull in the transverse direction and moves freely (or with friction) in the longitudinal direction.

Based on the above assumptions, a set of contact points and nodes closest to the contacts should be defined at the beginning of analysis. The detailed contact force formulations have been described in Appendix II. During numerical computations, these nodes are being monitored, and the contact forces are being computed and then transferred to the spar hull to solve the coupled dynamic equation of motion (Eq. 4-1).

Environment Modeling

Irregular wave with Jonswap spectrum, dynamic wind with API spectrum, and a current profile with velocities varied at different water levels are modeled. With the DeepCat model, wave, wind, and current are not required to be co-linear and are able to be input in any directions specified by the user.

Damped Mathieu Instability Evaluation

Damped Mathieu instability evaluation for spar design has been presented in Zhang et al. (2002). A brief review follows:

General Damped Mathieu Equation. A general damped Mathieu equation is shown as follows:

$$\ddot{x} + c\dot{x} + (a + b\cos\lambda)x = 0 \tag{4-2}$$

Equation 4-2 is similar to Eq. 4-1, but it has only one dimension unknown with time-varying stiffness ($a + b\cos\lambda$) and no external exciting force. By fixing c, zeros of infinite determinants can be found by specifying a (or b) and searching for the corresponding b (or a), which gives a set of results sufficiently close to zero. For engineering application, the damping is defined in a percentage of critical damping which is equal to $2\sqrt{a}$ in a sense of averaging the time-varying stiffness in Eq. 4-2.

Pitch/Heave Coupling. For a spar, the pitch stiffness (k_5) is a function of spar displacement (∇) and metacentric height (GM), denoted as ρg∇GM in still water. When the spar is heaving and the amplitude is assumed as u_3, then

$$GM_{new} = GM - \frac{1}{2}u_3 \tag{4-3}$$

$$\nabla_{new} = \nabla - A_w u_3 \tag{4-4}$$

$$k_{5new} = \rho g \nabla_{new} GM_{new} \tag{4-5}$$

where A_w is the spar water plan area.

Substituting Eqs. 4-3 and 4-4 into 4-5, gives

$$k_{5new} = k_5 - \frac{1}{2}\rho g(\nabla + 2A_w GM)u_3 + \frac{1}{2}\rho g A_w u_3^2 \tag{4-6}$$

It is clear from Eq. 4-6 that pitch stiffness is coupled with heave motion and is time-dependent. For simplification, the parametric excitation is assumed by one-term harmonics, $-1/2\,\nabla u_3 \cos(\omega t)$, where ω is heave motion frequency. Thus, the pitch motion can be written as follows:

$$(I_{55} + A_{55})\ddot{u}_5 + C\dot{u}_5 + \nabla(GM - 1/2\,u_3 \cos\omega t)u_5 = 0 \tag{4-7}$$

where

I_{55} and A_{55} = pitch moment inertia and added pitch moment inertia

u_5 and u_3 are pitch = heave motions

Compared to the general damped Mathieu equation,

$$a = \nabla \cdot GM/(I_{55} + A_{55})\omega^2 = (\omega_5/\omega)^2 \tag{4-8}$$

$$b = \nabla \cdot 0.5u_3/(I_{55} + A_{55})\omega^2 \tag{4-9}$$

$$c = C/(I_{55} + A_{55})\omega \tag{4-10}$$

where ω_5 is pitch natural frequency.

General Damped Stability Diagram

Damping coefficients equal to 0 percent, 1 percent, 5 percent, and 10 percent of critical damping are specified respectively. A stability diagram of the damped Mathieu equation (Eq. 4-2) is shown in Figure 4-2.

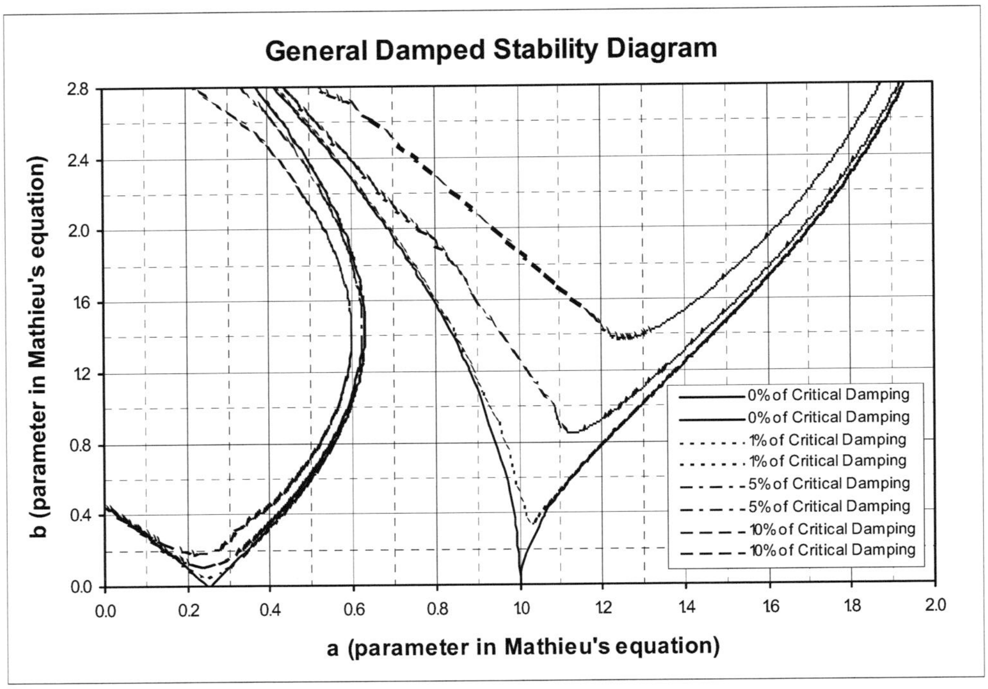

Fig. 4-2. General damped stability diagram

Damping Effects on Mathieu Instability

A few key notes drawn from Figure 4-2 are summarized as follows:

- Without damping ($b \approx 0$), instability occurs at $a = 0.25, 1, \ldots$.
- However, once damping is added to the system, the unstable zones are separated from the a-axis.
- Higher-order regions of instability are more sensitive to damping than low-order ones.
- Since the principal unstable zone ($a = 0.25$) is very important and less affected by damping, careful examination and evaluation of the principal unstable zone is needed.

Descriptions of Case Study

Spar Hull and Its Mooring/Riser Configurations

A spar model test setup in a wave basin is shown in Figure 4-3. The spar and its mooring/riser system configurations are illustrated in Figures 4-4 and 4-5 and summarized in Tables 4-1 through 4-3. There are a total of 14 mooring lines and 23 riser members, which have been modeled individually without performing any lumping or grouping. The layout of mooring lines and risers is shown in Figure 4-5.

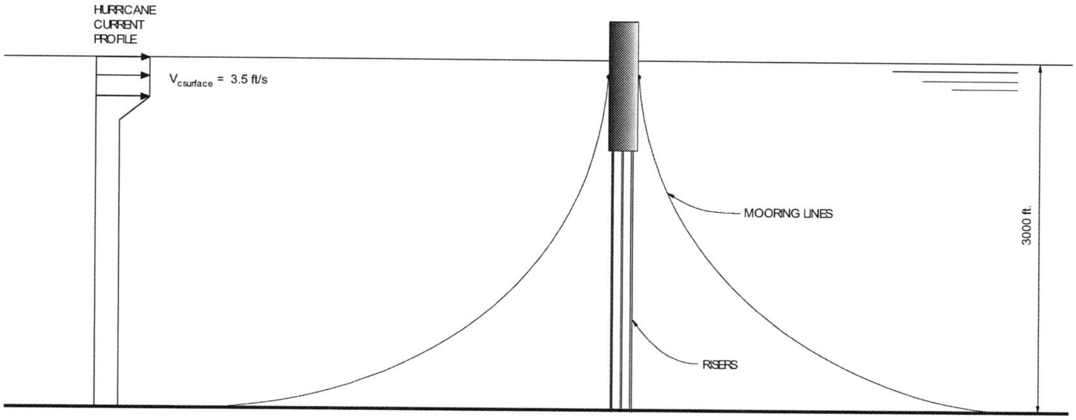

Fig. 4-3. Spar test setup in wave basin

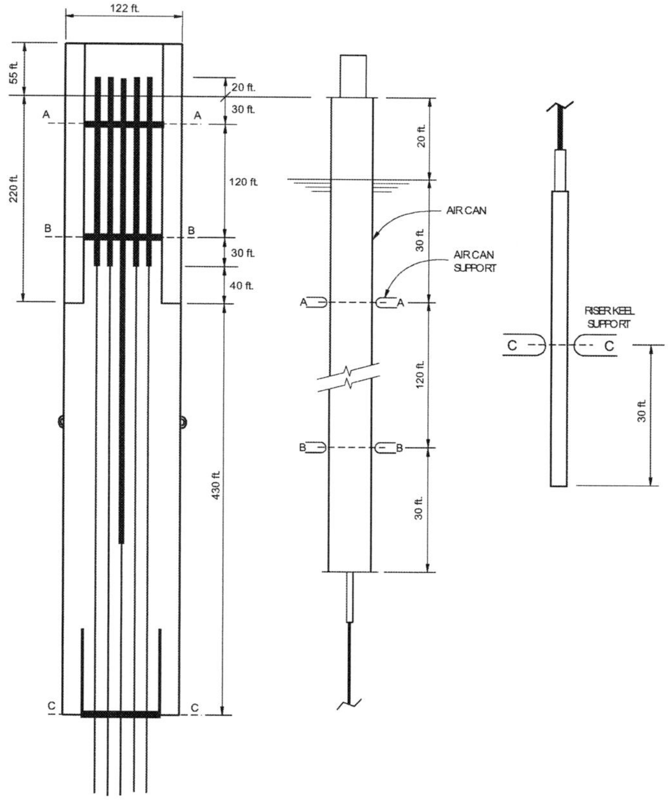

Fig. 4-4. Spar hull, air can, air can support and keel joint support

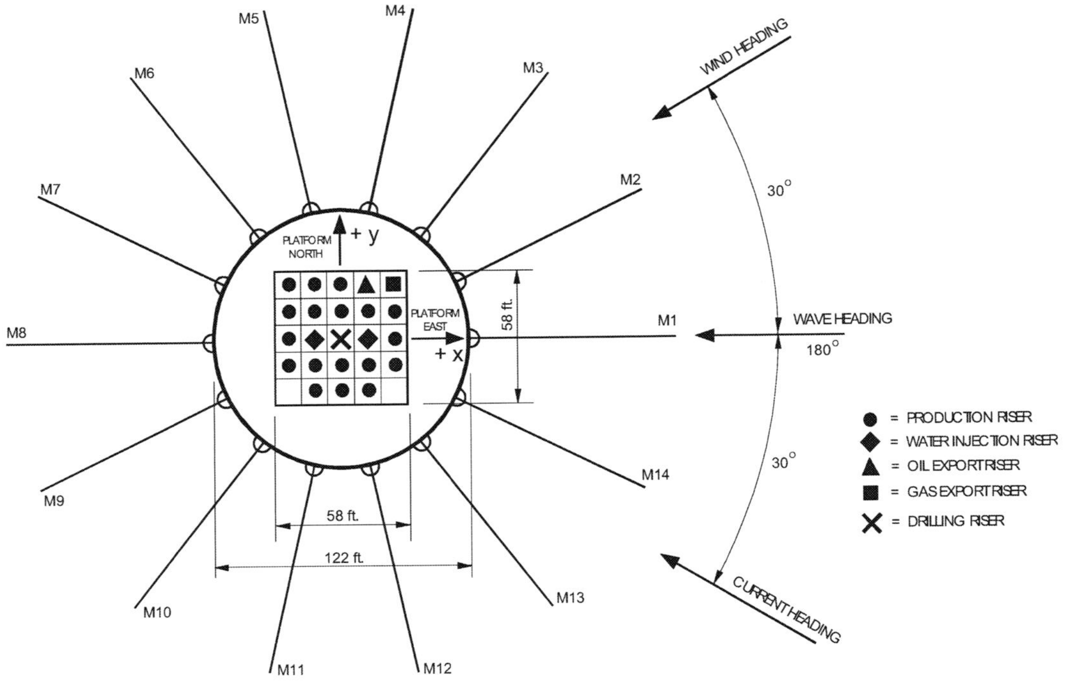

Fig. 4-5. Layout of mooring lines and risers and wave/wind/current directions

Table 4-1. Spar Hull Configuration

Design water depth (ft.)	3000
Draft (ft.)	650
Displacement (st)	243,310
Hard tank ht below MSL (ft.)	220
Hull diameter (m)	122
Central well bay square (25 slots)	58 ft. × 58 ft.

Table 4-2. Spar Mooring System Configurations

Mooring pattern	14-point taut-leg
Spread direction	Omni-direction
Platform section	250 ft. x 5-1/4 in., k4 studless chain
Riser section	3200 ft. x 5-3/8 in., heathed wire
Ground section	1150 ft. x 5-1/4 in., k4 studless chain

Table 4-3. Spar Riser Configurations

Riser Type	No.	Tension (kips) At base air can	Tension (kips) Just below keel	Tension (kips) At sea bed
Production (14.25 in)	19	527	473.4	157.14
Water injection (8.62 in)	2	324.4		221.9
Oil export (16.0 in)	1	442	390.7	132.2
Gas export (16.0 in)	1	214.2	199.4	61.15

Environmental Conditions

There are two sets of environmental conditions that have been specified. Set one is for spar dynamic responses in extreme conditions. Set two is for spar Mathieu instability evaluation.

Set one includes 100-year hurricane and 100-year loop current conditions in the Gulf of Mexico. The wave/wind/current headings are illustrated in Figure 4-5 with respect to the x-axis (Platform East). The 100-year hurricane wave Hs = 40.0 ft (12.19m), Tp = 14.0 s, Jonswap spectrum, overshoot $\gamma = 2.5$; associated wind speed is 134.8 ft./s (41.09 m) with API wind spectrum and surface current velocity is 3.51 ft./s (1.07 m/s). The 100-year loop current surface velocity is 6.14 ft./s (1.87 m/s); associated wave Hs = 10.2 ft. (3.11 m), Tp = 11.0 s, Jonswap spectrum, overshoot $\gamma = 2.0$; wind speed is 73.3 ft./s (22.34 m/s). The current heading is same as current shown in Figure 4-5 while wave and wind are in same direction and 90° counter-clockwise from current heading.

Set two consists of two long swell waves in West Africa. Swell one is the wave with Hs = 8.2 ft (2.5 m), Tp = 22.0 s, Jonswap spectrum, overshoot $\gamma = 6.0$. Swell two is the wave with Hs = 5.6 ft (1.71 m), Tp = 25.0 s, Jonswap spectrum, overshoot $\gamma = 6.0$.

Results and Discussions

Comprehensive physical model tests, including modeling all air cans, supporting frames, and keel joints have been carried out with scale 1:87. All 14 mooring lines were modeled full length without any truncation. Spar hull, mooring lines, and air can supported riser configurations are documented in Tables 4-1 through 4-3. The air cans and their supporting frames and keel joint details are shown in Figure 4-4. The layout of mooring lines and all risers and environment orientations are demonstrated in Figure 4-5.

Static Offset. The measured and simulated static offsets are shown in Figure 4-6. The measured and simulated static mooring line tensions of lines 1 and 8 shown in Figure 4-7 indicate that excellent agreement has been achieved.

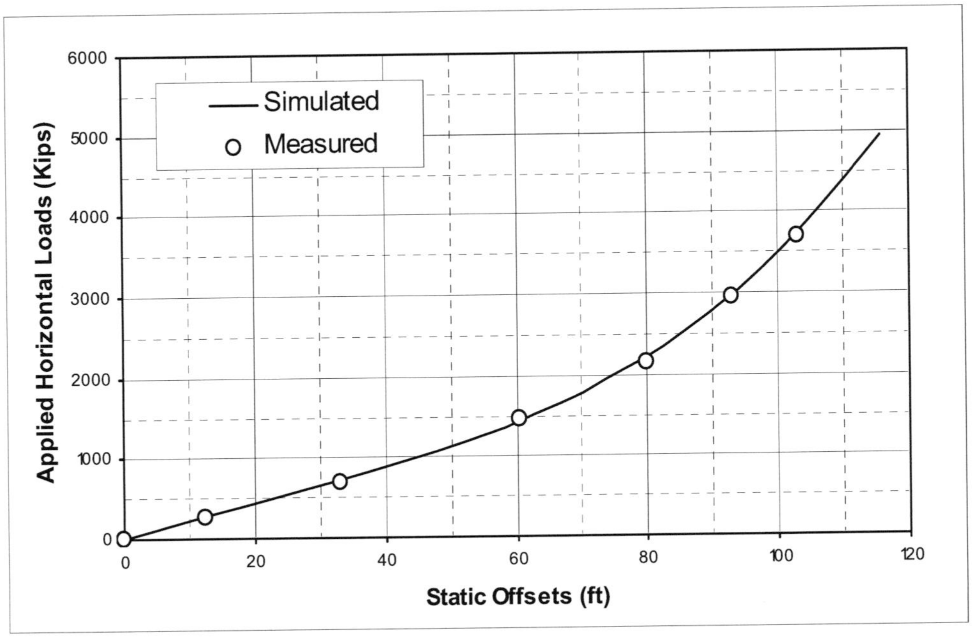

Fig. 4-6. Static offsets vs. horizontal loads (measured and simulated)

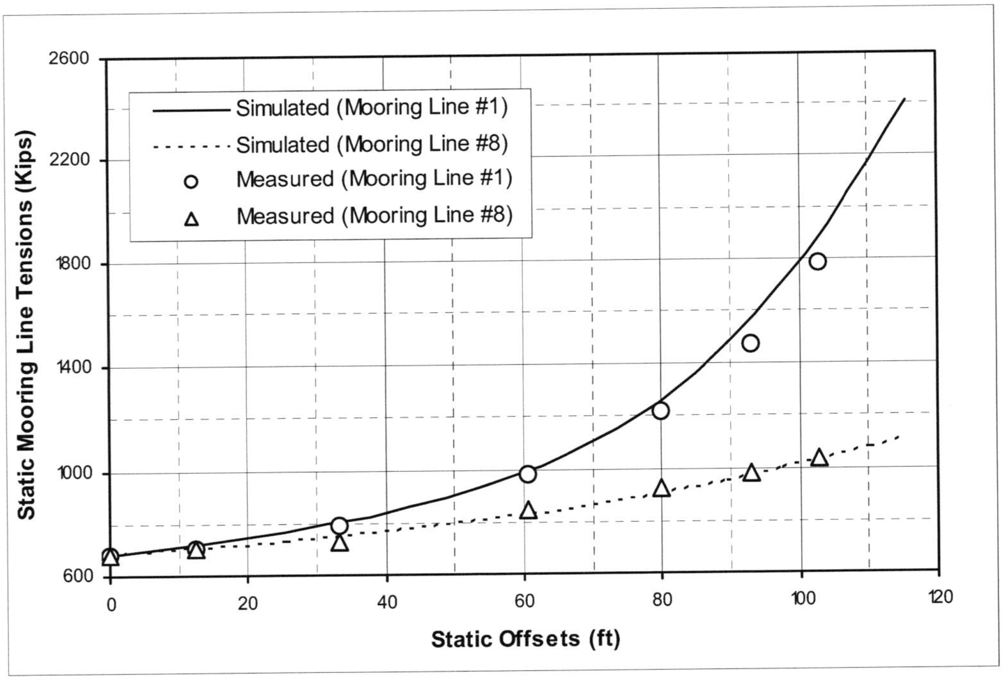

Fig. 4-7. Static line tensions vs. horizontal loads

Free-Decay

Table 4-4 summarizes the simulated and measured natural periods of surge/sway, heave, and pitch/roll. It is seen that agreements are very good.

Table 4-4. Natural Period Comparisons

Surge		Heave		Pitch	
Measured	Predicted	Measured	Predicted	Measured	Predicted
(sec)	(sec)	(sec)	(sec)	(sec)	(sec)
240	245	28	29	49	50

Damping Effects of Potential, Risers, and Mooring Lines

It is our intention to identify the damping contributions of mooring lines, risers, and wave diffraction damping (potential damping) to the spar heave motions. Figure 4-6 illustrates the damping effects due to potential, risers, and mooring lines. It is clear that wave potential damping is negligible. The damping from 23 risers is smaller than from 14 mooring lines. Friction damping due to risers will be a smaller portion for a truss spar because the truss spar system is highly damped by the heave plates. The mooring line damping is the major damping contributor to the classic spar system heave motions.

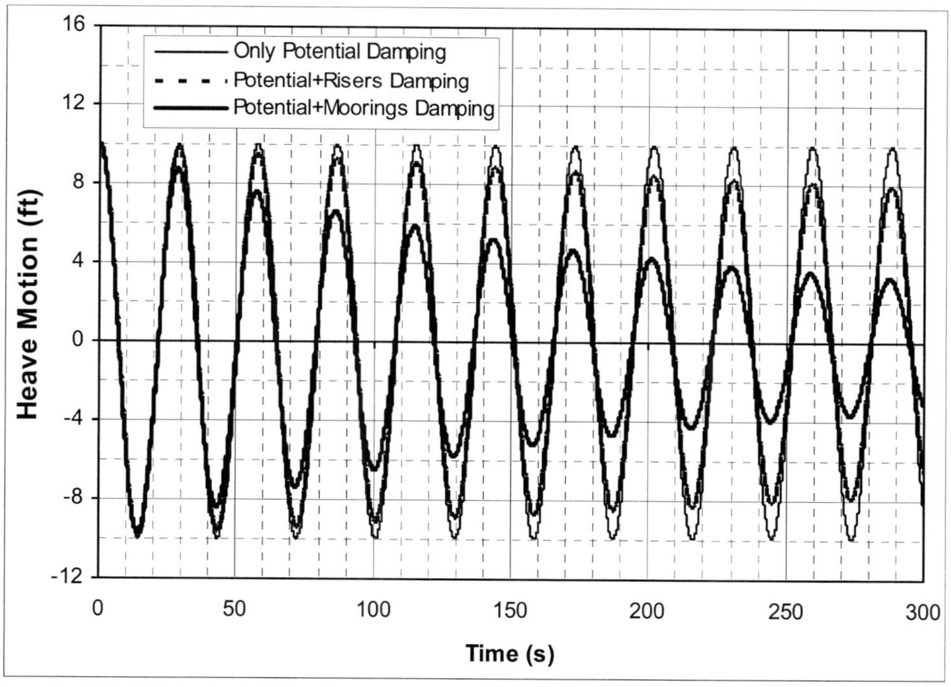

Fig. 4-8. Damping effects of potential, risers, and mooring lines

Spar Motions at CG, Measured vs. Simulated

Mean, standard deviation, and extreme values of simulated and measured motions at the spar CG are compared in Table 4-5.

- ***Surge Motion***
  The simulated mean is about 1.5 percent smaller than the measured. The standard deviation and extreme values are virtually the same.
- ***Heave Motion***

  The extreme values of measured and simulated heave motions are about 1.6 ft, which is small. This implies that it is possible to shorten the spar draft by about 60 ft. to 80 ft. and still maintain acceptable heave motion. This will reduce hull steel weight and improve hull transportation and fabrication.
- ***Pitch Motion***

  The simulated results of mean, standard deviation, and extreme value are about 28 percent, 15 percent, and 2.5 percent higher than the corresponding measured values respectively. The difference in mean values may be attributed to the inconsistency on the dynamic wind loads generated in the wave basin, the material property and geometry of the air cans, and the keel joints between the model test and the numerical modeling.

Figure 4-9 shows the comparison of pitch amplitude spectrum, simulated vs. measured. As shown, there is good agreement between simulated and measured around both the wave dominant energy region and the pitch natural period region, but there is overestimation around the surge natural period region.

Table 4-5. Measured and Simulated Spar Motions

	Surge		Heave		Pitch	
	Measured (ft)	Simulated (ft)	Measured (ft)	Simulated (ft)	Measured (deg)	Simulated (deg)
Mean	-63.3	-62.4	-0.66	-0.65	-1.20	-1.53
St. Dev.	6.9	7.0	0.36	0.30	0.80	0.92
Extreme	-85.3	-82.5	-1.87	-1.63	-4.90	-5.02

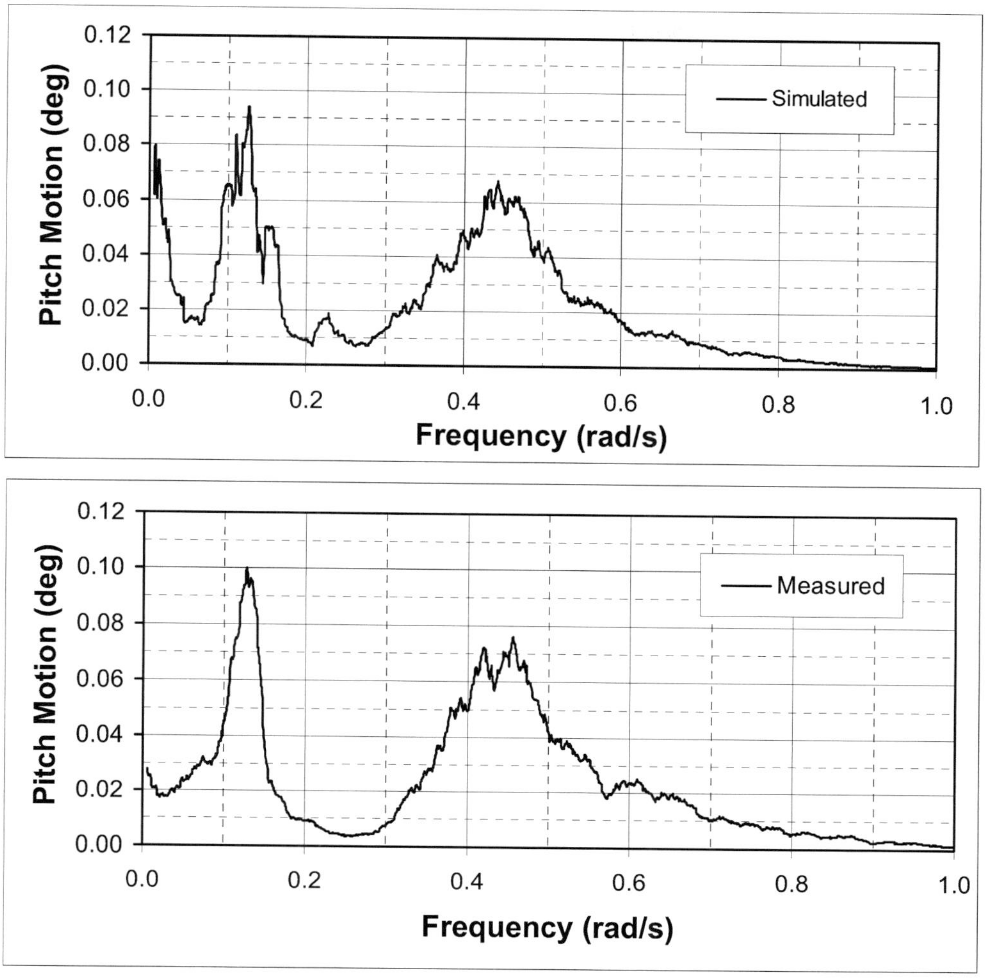

Fig. 4-9. Comparison of pitch amplitude spectrum, simulated vs. measured

Characteristics of Contact Forces

When the spar oscillates around the mean offset position, the hull will incline, and the air cans/riser keel joints will contact the supporting guide frames. Meanwhile, the roll/pitch and heave motions of the hull will result in the contact forces at the contact points.

Figure 4-10 shows a typical production riser contact force at the keel joint with respect to the pitch motions. Note that the phase difference between pitch and contact force is 180°. In other words, the induced moment of the contact force at keel is always opposite to the pitch motion. This means the pitch motion will be less once the restoring moment is taken into account. From our calculation, the maximum contact force on one production riser is about 45 kips. Because the arm of resisting moment is very long (about 295 ft. to system CG), the resisting moment due to single production riser is about 13,275 kips-ft. Considering that there are a total of 23 risers, the total resisting moments are very large, and the impact on the pitch motion of the hull cannot be ignored.

Figure 4-11 illustrates contact force amplitude spectrum at keel joint. There are several peaks due to the coupling effects of surge, heave, and pitch motions. There are two major peaks, which are (1) around the wave dominant engery region, and (2) around the pitch natural period region. Figures 4-10 and 4-11, clearly reveal that contact forces are strongly coupled with pitch motions.

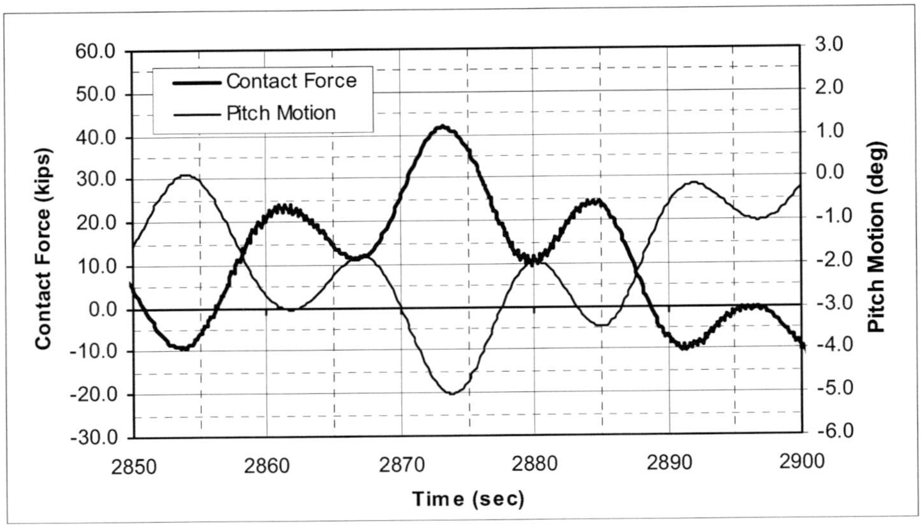

Fig. 4-10. Contact force time series at keel joint

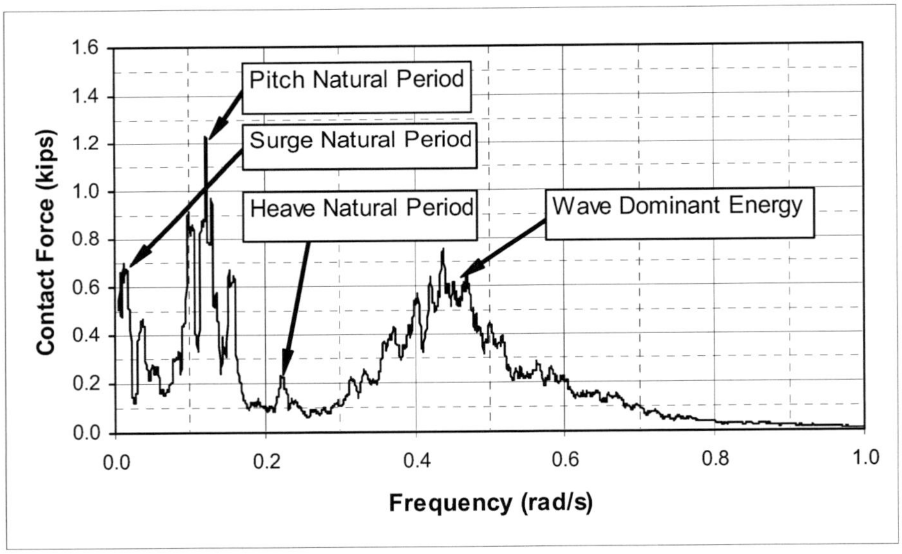

Fig. 4-11. Contact force amplitude spectrum

Characteristics of Mooring Line Dynamics

Mean, standard deviation, and extreme values of simulated and measured mooring line tensions at the top are compared in Table 4-6. As shown in Figure 4-5, mooring lines #1, #13, and #14 are the most loaded lines for the given environmental condition.

- ***Mooring line #1.*** *The simulated res*ults of mean, standard deviation, and extreme values are about 2.6 percent, 12 percent, and 6.8 percent higher than the corresponding measured results, respectively.
- ***Mooring line #13.*** *The simulated res*ults of mean and extreme values are about 4.6 percent and 1.3 percent higher than the measured results, respectively. The standard deviation is about 6.7 percent lower than measured.
- ***Mooring line #14.*** *The simulated resu*lts of mean, standard deviation, and extreme values are about 5.7 percent, 6.6 percent, and 8.3 percent higher than the measured results, respectively.

Table 4-6. Measured and Simulated Mooring Line Tensions

	Mooring Line #1		Mooring Line #13		Mooring Line #14	
	Measured (kips)	Simulated (kips)	Measured (kips)	Simulated (kips)	Measured (kips)	Simulated (kips)
Mean	1012	1038	859	898	967	1021
St. Dev.	88	98	54	50	81	87
Extreme	1377	1470	1073	1087	1315	1424

Figure 4-12 shows the top dynamic tension of the most loaded mooring line #1 vs. the hull surge motion. It is clearly shown that the top tension of mooring line #1 reaches maximum, while the surge motion goes to extreme.

Figure 4-13 illustrates mooring line #1 top dynamic tension amplitude spectrum, simulated vs. measured. There are two major peaks, which are (1) around the wave dominant engery region and (2) around the surge natural period region. The results shows good agreement between simulated and measured around both wave dominant energy regions, but there is overestimation around the surge natural period region.

Figures 4-12 and 4-13 clearly reveal that mooring line dynamic tensions are strongly related to surge/sway motions.

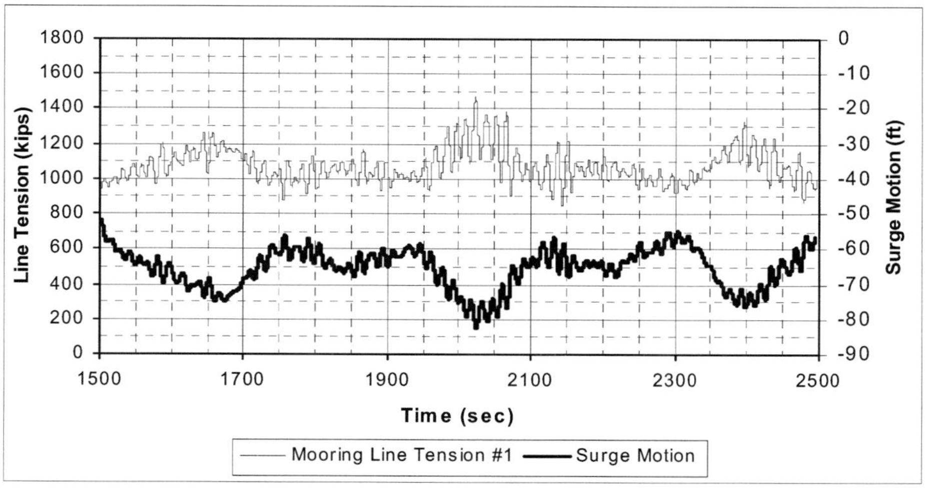

Fig. 4-12. Mooring line #1 top dynamic tension vs. surge motion

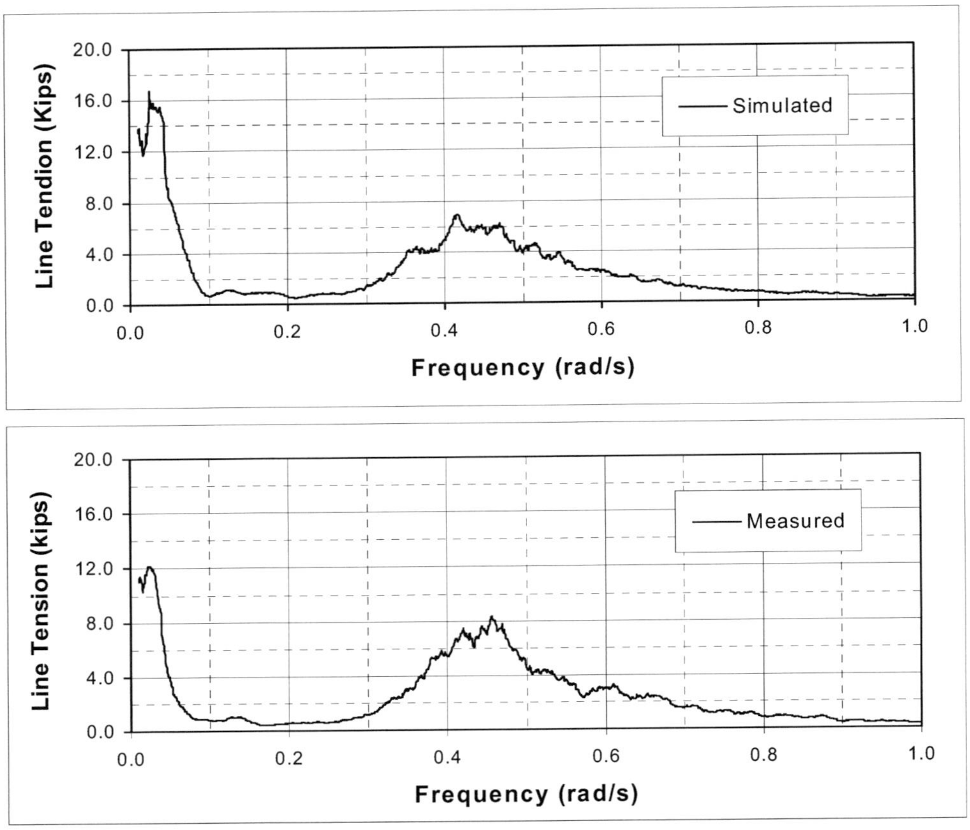

Fig. 4-13. Mooring line tension amplitude spectrum, simulated vs. measured

Characteristics of Vortex Induced Motion (VIM)

DeepCat could not simulate the mechanics of vortex induced motions. It has been reported that the indirect method, by imposing forcing functions to excite spar motion in a transverse direction to explore the VIM, affects mooring lines and risers design. This alternative method has not been implemented in the DeepCat. Figure 4-14 compares the measured spar trajectory motions at CG between those with and without strakes in currents. In this figure, only 100-year loop currents were generated; no associated wave and wind were applied. Classic figure 8 motions show for spar without strakes while the transverse motions were suppressed considerably with the strakes. The left plot in Figure 4-14 clearly reveals strong VIM exists.

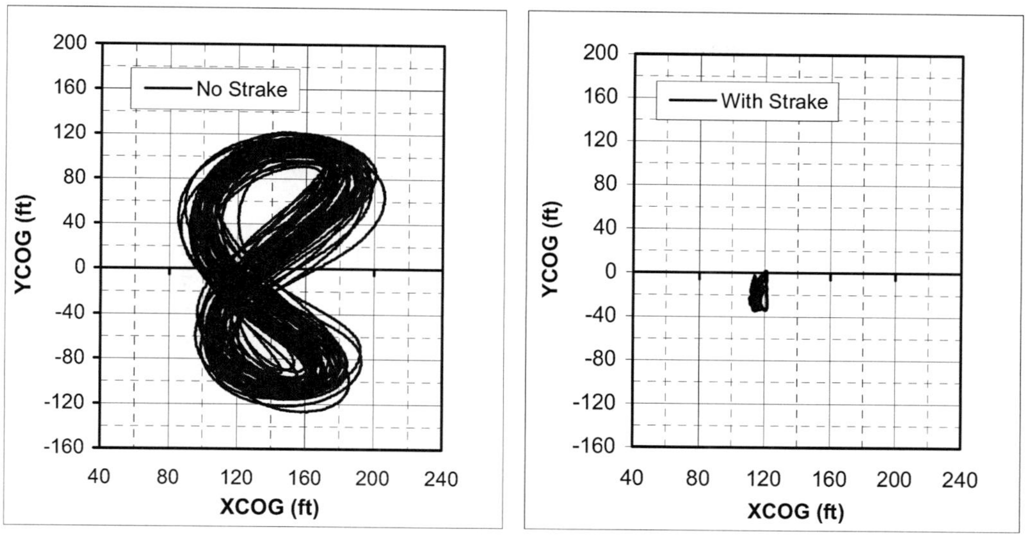

Fig. 4-14. Comparison of the measured spar motions at CG in X-Y plan between those with and without stakes in the 100-year loop currents (current only)

Affects of Spar VIM on Mooring Line Tension

Figure 4-15 shows the comparison of the measured most loaded mooring line tensions between with and without strakes. As presented in Figure 4-14, the spar has much stronger VIM in loop current without strakes than with strakes. The affects of strong VIM on mooring line tensions are dramatic as illustrated in Figure 4-15. The mooring line will break without strakes to suppress VIM.

Affects of Spar VIM on Riser Tension and Stroke

Figures 4-16 and 4-17 illustrate the comparisons of the measured production riser bottom tensions and vertical motions with and without strakes in 100-year loop currents (current only) respectively. The impacts of strong VIM on riser tension and stroke are significant as presented in Figures 4-16 and 4-17.

Figures 4-15 to 4-17 clearly demonstrate the importance of suppressing VIM on the mooring lines, in risers design, and for overall project initial cost (CAPEX) and operation cost (OPEX).

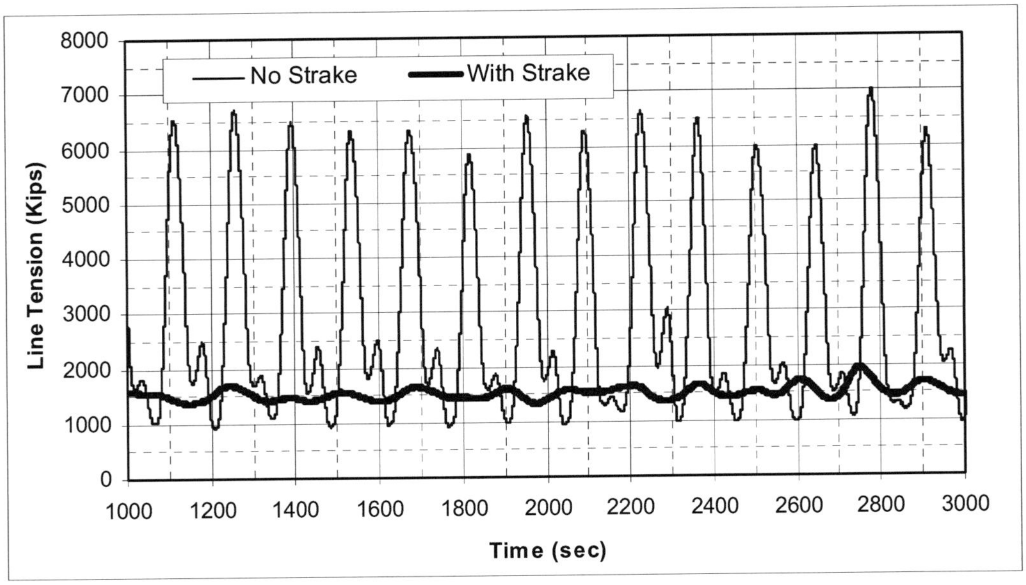

Fig. 4-15. Comparison of the measured most loaded mooring line tensions with and without stakes in 100-year loop currents (current only)

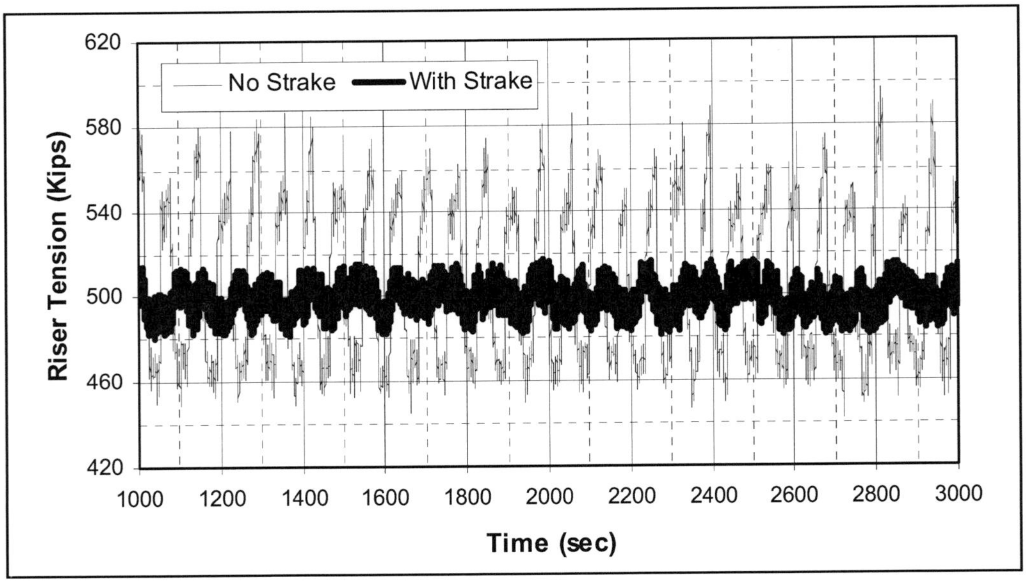

Fig. 4-16. Comparison of the measured production riser bottom tensions with and without stakes in 100-year loop currents (current only)

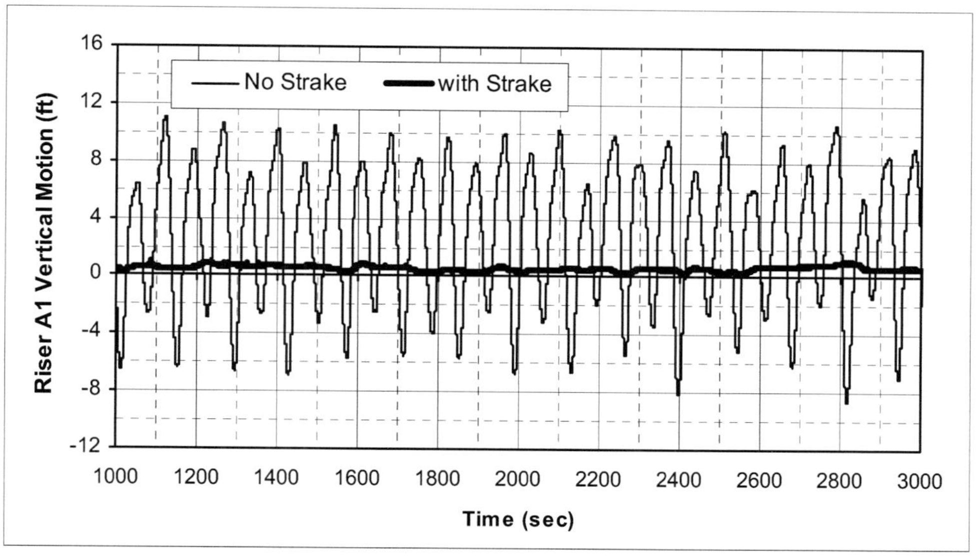

Fig. 4-17. Comparison of the measured production riser bottom vertical motions with and without stakes in 100-year loop currents (current only)

Damping Effects on Pitch/Heave Coupling

Figure 4-18 shows the normalized measured heave amplitude spectrum without strakes and heave free decay in still water vs. heave amplitude spectrum with strakes and free decay in current. The sharp peak around the pitch natural period clearly indicates that there is pitch/heave coupling when there are no strakes and free decay in still water, but the sharp peak around the pitch natural period disappears with strakes and free decay in the current. Figure 4-18 reveals the effects of damping on pitch/heave coupling.

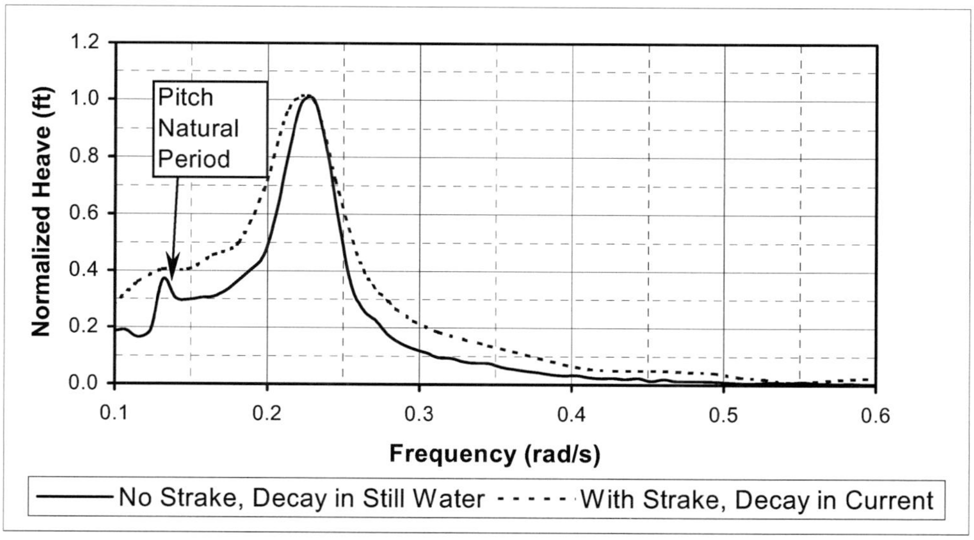

Fig. 4-18. Normalized heave amplitude spectrum, no strakes in still water vs. with strakes in current (current surface velocity 6.14 ft./sec.)

Damping Effects on Heave Motions in Long Swell Waves

It has been reported that there are peak period 25-second swell waves offshore West Africa and in the North Sea. This creates a great design challenge, as ways must be found to reduce heave motions to an acceptable level without substantially increasing cost or inducing instability.

Figure 4-19 shows spar heave motion RAOs computed by WAMIT with 1 percent, 2 percent, and 3 percent of heave critical damping. It is seen that heave amplitude around the natural period region is very sensitive to damping. Two West African long swell wave spectral densities are also given in Figure 4-19.

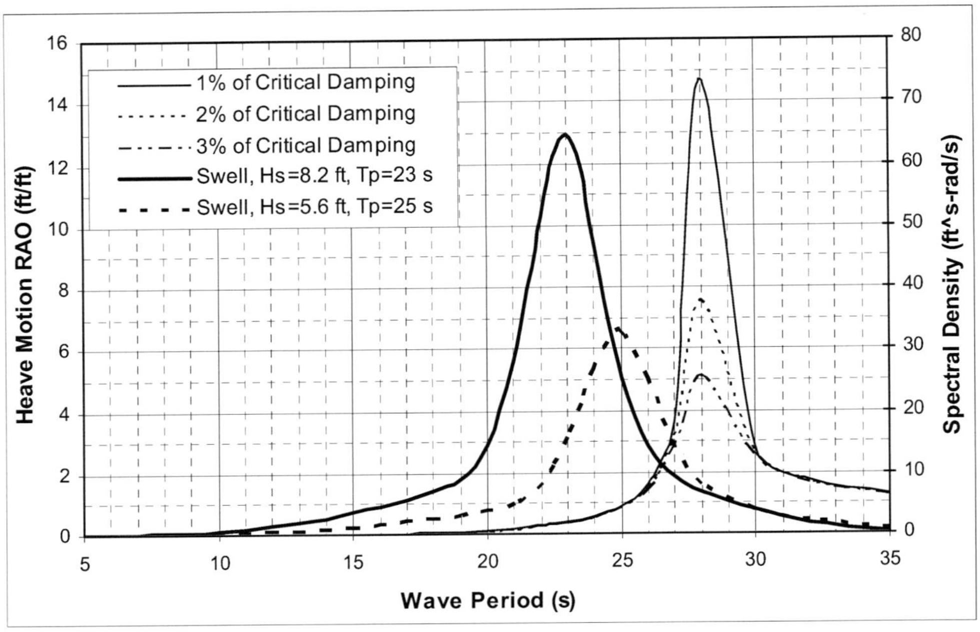

Fig. 4-19. Conventional spar heave motion RAOs

Table 4-7 summarizes the maximum heave motions in two long swell waves with the wave frequency component only. There also may be a low-frequency heave motion component, which is not included in Table 4-7.

Table 4-7. Maximum Heave Motion in Long Swell Waves

	1% of Heave Critical Damping	2% of Heave Critical Damping	3% of Heave Critical Damping
Swell, Hs=8.2 ft, Tp=23 s, Gamma=6.0	17.9	10.6	8.2
Swell, Hs=5.6 ft, Tp=25 s, Gamma=6.0	19.5	11.6	9.0

Assessments of a Spar's Mathieu Instability in Principal Unstable Zone

As shown in Figure 4-2, there are two sets of family curves to define boundaries of an unstable zone varying with damping. For example, if no damping, a = 0.25, then the area above the solid line is defined as an unstable zone. The first set of family curves defined as an unstable zone is called the principal unstable zone, while the second set of family curves defined as an unstable zone is called the secondary unstable zone. It is seen that the secondary unstable zone is much more sensitive to damping than the principal unstable zone.

On the basis of Table 4-4 and Eq. 4-8, spar pitch and heave natural frequency do not satisfy $(\omega_5/\omega_3)^2 = 0.25$, where ω_5 and ω_3 are pitch and heave natural frequencies, respectively. But this does not mean that there is no problem. If there is an incidence of a regular wave with a period of 25 seconds, the spar heave motion will consist of the following four frequencies: a wave period incidence of 25 seconds, a heave natural period of 28 seconds, a difference period of 233 seconds, and a sum period of 13.2 seconds. In this condition, the difference and sum periods are not a concern. Figure 4-2 shows that if $(\omega_5/\omega_i)^2 = 0.25$, b is approximately equal to zero (no damping) where ω_i is one of heave motion frequencies, and any small heave motion may excite the Mathieu instability.

On the basis of Eqs. 4-8 to 4-10 and Figure 4-2, maximum allowable heave motions around the principal unstable zone associated with two long swell waves are given in Table 4-8. From this table, two key points are summarized as follows:

- For a swell wave with Tp = 23.0 s, there is no instability concern if heave damping is equal to or larger than 1 percent of heave critical damping.
- For a swell wave with Tp = 25.0 s, there is no instability concern if heave damping is equal to or larger than 2 percent of the heave critical damping and the pitch damping is equal to or larger than 1 percent of the pitch critical damping. Otherwise, an instability problem could possibly occur. Further extensive examinations and physical model tests would be recommended.

Table 4-8. Comparison of Maximum Allowable Heave Motion and Maximum Heave Motion Varying with Heave and Pitch Damping Around Principle Zone

	0% of Pitch Critical Damping	1% of Pitch Critical Damping	5% of Pitch Critical Damping	10% of Pitch Critical Damping
Swell, Hs=8.2 ft, Tp=23 s, Gamma=6.0				
Maximum allowable heave motion	22.0	22.6	32.8	49.2
Maximum heave motion				
1% of heave critical damping, 17.9 ft.	No	No	No	No
2% of heave critical damping, 10.6 ft.	No	No	No	No
3% of heave critical damping, 8.2 ft.	No	No	No	No
Swell, Hs=5.6 ft, Tp=25 s, Gamma=6.0				
Maximum allowable heave motion	7.2	14.8	26.2	45.9
Maximum heave motion				
1% of heave critical damping, 19.5 ft.	Yes	Yes	No	No
2% of heave critical damping, 11.6 ft.	Yes	No	No	No
3% of heave critical damping, 9.0 ft.	Yes	No	No	No

Note:
"Yes" means maximum heave motion is larger than maximum allowable heave motion.
"No" means maximum heave motion is less than maximum allowable heave motion.

Assessments of a Spar's Mathieu Instability in Secondary Unstable Zone

If $(\omega_5/\omega_i)^2 = 1.0$ satisfies, the secondary unstable zone may be triggered, where ω_i is one of the heave motion frequencies. This seems to be impossible because the longest period wave observed is less than 30 seconds. However, it still can be triggered due to the difference in frequency excitation as shown in this section. If the incident regular wave has a period of 17.82 seconds, the spar heave motion will consist of four frequencies, which have incident wave periods of 17.82 seconds, heave natural periods of 28 seconds, a difference period of 49 seconds, and a sum period of 10.89 seconds. Thus, the condition of $(\omega_5/\omega_i)^2 = 1.0$ can be satisfied. From Figure 4-2, b is approximately equal to zero in the absence of damping. This means any small heave motion may generate Mathieu instability.

Maximum allowable heave motions around the secondary unstable zone are summarized in Table 4-9. This table shows that maximum allowable heave motions increase rapidly with the pitch damping increasing. Therefore, pitch damping percentage in the system determines whether Mathieu instability due to heave difference frequency excitation will occur or not.

Table 4.9. Maximum Allowable Heave Motion Varying with Pitch Damping Around Second Unstable Zone

	0% of Pitch Critical Damping	1% of Pitch Critical Damping	5% of Pitch Critical Damping	10% of Pitch Critical Damping
	ft	ft	ft	ft
Heave Difference Frequency Effect	3.3	21.7	62.3	85.3

Conclusions and Recommendations

Dynamic responses in extreme conditions and Mathieu instability evaluation of a conventional spar in 3000 ft. have been performed. Conclusions are drawn as follows:

The coupling effects of risers and supporting guide frames on hull motions and mooring line dynamics have been identified. It has been found that the resisting moment induced by the contact forces has significant impact on both the pitch/roll motion and the mooring line tensions. The results show that both pitch/roll motion and the mooring line tensions are reduced when the effects of contact forces on the supporting guide frames are taken into consideration.

The analytical results are further validated by the physical model test results. Good agreement between the simulated and the measured have been achieved. The comparisons indicate the present enhanced DeepCat program can simulate the dynamic response of spar accurately with the presence of non-collinear wave, wind, and current.

DeepCat could simulate the mechanics of vortex induced motions. It has been reported that the indirect method, by imposing forcing functions to excite spar motion in transverse direction to explore the VIM, affects the design of mooring lines and risers. This alternative method has not been implemented in the DeepCat. The measured VIM and its impacts on mooring line tensions, riser tensions, and strokes have been presented. It clearly demonstrates the importance of suppressing VIM on the mooring lines, risers design, and overall project initial cost (CAPEX) and operation cost (OPEX).

A general stability diagram has been generated. It has been found that high-order regions of instability are more sensitive to the damping than low-order ones. The regions of the principal and secondary unstable zones are the major concerns for FPS design. It has been demonstrated that there are damping effects on suppressing Mathieu instability.

The major contributions to heave damping are the mooring lines and risers/supporting guide frames inside the moonpool. Coupled analyses including these effects are needed to estimate heave damping accurately. Without accurate and consistent heave damping estimation, misleading conclusions may be drawn on Mathieu instability.

Through the investigations of spar hull/mooring/riser coupled dynamic analyses and spar VIM in loop currents and Mathieu instability evaluations, recommendations are made as follows:

Smaller heave and pitch/roll motions in reality will lead to reduced hull draft and solid ballast in the keel tank. These reductions have a significant influence on hull sizing, hull fabrication, and transportation. It is recommended that spar hull/mooring/riser coupled dynamic analysis be performed, including contact forces to prevent unnecessary conservatism and to improve the economics of the field development.

It is important to investigate the affects of spar VIM on mooring line and riser strength and fatigue design. Suppressing VIM efficiently has significant economic impacts on spar CAPEX and OPEX.

Applying the following aspects can help avoid the principal unstable problem in the design:

- Increase the pitch natural period, and make it higher than 60 seconds.
- Increase the heave natural period, and make it higher than 30 seconds to reduce the excitation value such as b in long swell waves.
- Increase heave and pitch damping to suppress pitch/heave coupling.

Acknowledgments

This article was written in July 2002. At that time, the author was the principal engineer of ABB Lummus Global, Deepwater System Division. The article has been revised slightly to trace the status of key contributors. The author is grateful to Dr. Wei Ma and Ming-Yao Lee, formerly with ABS and currently with Chevron, for their contributions during early phase DeepCat program development. The author wishes to thank his colleagues, especially to Dr. Xinyu Zhang, formerly with ABB and currently with DNV, for his assistance in contact force computations and Dr. Libang Zhang, formerly with ABB and currently with Technip, for his assistance in Mathieu instability evaluation. The author would like to express appreciation to Mr. John Chianis, formerly with ABB and currently with HOE, for his support and constructive comments.

Appendix I: Line Dynamics

The equations of motion for the immersed line are

$$\mathbf{M}\ddot{\mathbf{r}} + (\mathbf{B}\mathbf{r}'')'' - (\tilde{\lambda}\mathbf{r}')' = \mathbf{q}_f^{(1)}(s) + \mathbf{q}_f^{(2)}(s) + \mathbf{q}_f^{(3)}(s) \tag{A4-1}$$

where $\mathbf{q}_f^{(1)}(s)$, $\mathbf{q}_f^{(2)}(s)$ and $\mathbf{q}_f^{(3)}(s)$ represent inertia force term, drag force term and Froude-Krylov force term respectively, in which

$$\mathbf{q}_f^{(1)}(s) = \rho_f A_f (1 + C_m \mathbf{N})\dot{\mathbf{v}}_f \tag{A4-2}$$

$$\mathbf{q}_f^{(2)}(s) = \frac{1}{2}\rho_f D_f \{\mathbf{N}(\mathbf{v}_f - \mathbf{v})\}|\mathbf{N}(\mathbf{v}_f - \mathbf{v})| \tag{A4-3}$$

$$\mathbf{q}_f^{(3)}(s) = g[\rho_f A_f - \rho_i A_i - \rho_t g(A_f - A_i)]\mathbf{e}_y \tag{A4-4}$$

In eq A4-2,

$$\mathrm{M} = (\rho_t(A_f - A_i) + \rho_i A_i)\mathbf{I} + \rho_f A_f C_m \mathbf{N} \tag{A4-5}$$

In Eq. A4-6, M is called effective mass matrix; A_f and A_i are the outer and inner cross-section area of the line respectively; B = EI, is the bending stiffness of the line; I is the identity matrix; N is an operator that yields normal to the line; and r(s,t) denotes a vector distance from the origin of the coordinate system as a function of s, the arc length along the line, and varied with the time, t

$$\tilde{\lambda} = (T + \rho_f A_f - \rho_i A_i) - B\kappa^2 \tag{A4-6}$$

In Eq. A4-7, the term in the parentheses is called the effective tension; κ is local curvature; T is the tension; C_d and C_m are the drag and inertia mass coefficients respectively; $\dot{\mathbf{v}}_f$ and $\mathbf{v}_f$ are the fluid acceleration and velocity at s respectively; $\mathbf{v}$ is the velocity of the line at s; D_f is the local diameter of the line; ρ_f, ρ_i, and ρ_t are the mass density of the sea water, the liquid inside the line, and the mass density of the line, respectively; and $\mathbf{e}_y$ is the unit vector in y direction.

Considering the extendibility of the line, one obtains

$$r' \times r' = 1 + 2\frac{T}{EA_t} \tag{A4-7}$$

After discretizing the above vector governing equations into algebraic equations by the Galerkin's method, the original vectored equations are solved using second-order Adams-Moulton integration algorithm (Ma and Webster 1994).

Appendix II: Contact Forces

Multiplying both sides of Eq. A4-1 in Appendix I with the shape function $a_i(s)$ and integrating it with respect to s from 0 to L for a segment (or an element) of the line with length L:

$$\int_0^L \{M\ddot{r} + (Br'')'' - (\tilde{\lambda}r')' - q\}a_i(s)ds = 0 \tag{A4-8}$$

Integrating the above equation, results in

$$\int_0^L \{M\ddot{r}a_i(s) + Br''a_i''(s) + \tilde{\lambda}r'a_i'(s) - qa_i(s)\}ds$$
$$= Br''a_i'\,|_0^L + \{\tilde{\lambda}r' - (Br'')'\}a_i\,|_0^L \tag{A4-9}$$

The first term on the right-hand side of Eq. A4-8 is the bending moment at the end of the element. The second boundary term is the force at the end. When the support guide acts on the node, the contact force can be treated as a concentrated load. Then the second term on the right-hand side of Eq. A4-8, from which the contact force can be calculated, will not vanish.

References

Garrett, D.L. (1982). "Dynamic Analysis of Slender Rods," OMAE 1982.

Gupta, H., Finn, L.D. and Weaver, T.O. (2000). "Effects of spar Coupled Analysis," Proc OTC 12082, Houston, TX, USA.

Haslum, H.A. and Faltinsen, O.M. (1999). "Alternative Shape of spar Platforms for Use in Hostile Areas," OTC 10953.

Kim, M.H., Roesset, J.M., and Zhang, J. (1997). "Nonlinear Dynamic Analysis Methods for Spar Platforms," Proc SNAME Conf (Gulf Section), Houston, TX, USA.

Kim, M.H., Ran, Z., and Zheng, W, (2001). "Hull/Mooring Coupled Dynamic Analysis of a Truss spar in Time Domain," Int J Offshore and Polar Eng, ISOPE, Vol 11, No 1, pp 42-54.

Ma, W., Lee, M.Y., Zou, J., and Huang, E.W. (2000). "Deepwater Nonlinear Coupled Analysis Tool," Proc OTC 12085, Houston, TX, USA.

Ma, W. and Webster, W.C. (1994). "An Analytical Approach to Cable Dynamics: Theory and User Manual," Sea Grant Project R/OE-26.

Paulling, J.R. and Webster, W.C. (1986). "A Consistent, Large-Amplitude Analysis of the Coupled Response of A TLP and Tendon System," OMAE 1986.

Zhang, L.B., Zou, J., and Huang, E.W. (2002). "Mathieu Instability Evaluation for DDCV/spar and TLP Tendon Design," Proc of the 11th Offshore Symposium, SNAME, Houston, TX, USA.

Zhang, X.Y., and Zou, J. (2002). "Coupled Effects of Risers/Supporting Guide Frames on spar Responses" ISOPE.

Chapter 5: Coupled Analysis of a Spar Using Slender-Body Formulas

By Jun Zhang and Yu Ding, Ocean Engineering Program, Department of Civil Engineering, Texas A&M University

Abstract

The natural frequencies of a typical spar are far below the spectral peak frequency of storm ocean waves. When deployed in ultra-deep water (deeper than 10,000 ft.), the hull of a spar is likely positioned by an integrated polyester mooring system. Because the global motion of a spar is dominated by slow-drift motions, accurate prediction of slow-drift (low-frequency) motions of a spar in stormy seas is a critical design issue and can only be achieved using numerical schemes based on nonlinear wave theory and accounting the interaction between the hull and its mooring system. A numerical code, known as COUPLE, was developed recently for computing 6 degrees-of-freedom (6DOF) motions of a moored floating structure dynamically interacting with its mooring/riser/tendon system. This paper describes the principles and efficacy of COUPLE for simulating the global motion of a spar positioned by an integrated polyester mooring system in ultra-deep water.

Introduction

The natural frequencies of a spar are usually much smaller than the spectral peak frequency of stormy ocean waves. Consequently, the responses of a spar at the wave frequency (WF) are small but relatively large near its natural frequencies even though incident waves involve insignificant energy at these low frequencies (LF). This indicates that nonlinear wave mechanics play an important role in the responses of a spar. Accurate prediction of slow-drift (LF) motions of a spar is a critical design issue and cannot be achieved using numerical schemes based on linear wave theory. Numerical schemes, such as WAMIT (Lee 1995), solve first- and second-order diffraction potentials using a panel method and then derive the wave loads by integrating pressure on the wetted surface of a floating structure at its mean position. Because the ratio of the diameter of a spar to a typical spectrum-peak wavelength is relatively small, the slender body approximation is valid. This leads to a relatively simple alternative, the use of the Morison equation to compute both potential and drag forces induced by waves. Accurate wave kinematics used in the Morison equation is crucial to achieve accurate prediction of wave loads, which can be accomplished using a deterministic hybrid wave model (HWM). An HWM considers nonlinear wave interactions in an irregular wave field up to second order of wave steepness in both decomposition and superposition of an irregular wave field and is able to accurately predict incident wave kinematics, including the contribution from nonlinear difference-frequency interactions (Zhang et. al. 1996, 1999), which are crucial to the slow-drift motions of a spar.

In addition to wave loads, current, and wind loads on a spar also result in significant slow-drift motions and are usually computed using empirical formulations. At present, the variation in the current velocity at the scale of the natural periods of a spar has not been considered in the simulation because of the lack of sufficient field measurements about this variation. The total loads on a spar together with the restraining forces resulting from its mooring and riser systems dictate its global motion. Depending on whether or not the dynamic forces in mooring lines and risers are considered in the computation of the global motions, the related analysis is coined as coupled or quasi-static analysis, respectively. Although restraining forces contributed from the dynamics of a mooring or riser system are not significant in comparison with wave loads in the WF range, they play an important role in damping or sometimes amplifying the responses of a spar near its natural frequencies, that is, the LF range. Therefore, it is crucial to employ a coupled analysis in computing the global motions of a spar.

A numerical code developed recently, known as COUPLE, is especially effective and relatively simple in predicting dynamic interactions between a spar and its mooring/riser systems. Initially, it was developed for computing the 3-degrees-of-freedom (3DOF) motions of a spar positioned by taut mooring lines using a quasi-static analysis (Cao and Zhang 1997). Later it was extended to allow for dynamical interaction between a spar and its mooring system to quantify the damping effects of a mooring system on the slow-drift motion of a spar (Chen et. al. 1999). More recently it was extended to allow for 6DOF motions of a moored structure. COUPLE consists of two basic computational parts—one for computing the dynamics of a mooring/tendon/riser system and the other for the wave/current/wind loads on a moored floating structure (hull). The two independent codes are coupled by matching the forces and displacements of a mooring/tendon/riser system and the hull at their joints following prescribed connection conditions. The code for computing dynamics of the mooring/riser system is based on a slender-body assumption and employs a nonlinear finite element method (FEM), known as CABLE3D (Ma and Webster 1994). The computation in the original CABLE3D assumes infinitesimal elongation of a slender rod. Because large elongation slender components, such as springs and polyester ropes are often used in a model test and a prototype mooring system, CABLE3D was extended to allow for large elongation in a mooring line to achieve accurate simulation (Chen et al. 2000). The computation of nonlinear wave forces on a floating structure is accomplished by using either a second-order diffraction wave theory, such as WAMIT, and/or the Morison equation. In the case of a spar, both potential and drag wave loads are computed using the Morison equation.

Equations for Met-Ocean Loads on the Hull

The total met-ocean environmental loads on an offshore structure can be divided into three major parts according to their origins denoted by the subscripts.

$$F = F_{Wave} + F_{Wind} + F_{Current}$$

The hull of a classical spar or the upper portion of a truss spar is virtually a cylinder. In using the Morison equation to compute wave and current loads, the normal force per unit length on a cylinder of uniform diameter D is given by,

$$\begin{aligned} dF_n &= (1 + C_m)\rho_f \frac{\pi}{4} D^2 (a_f)_n - C_m \rho_f \frac{\pi}{4} D^2 a_n \\ &+ \frac{1}{2}\rho_f C_D D \left|(v_f)_n - v_n\right| \left[(v_f)_n - v_n\right] \end{aligned} \tag{5-1}$$

where C_m is the added mass coefficient, C_D the drag coefficient, ρ_f the density of water, v_f and a_f water particle velocity and acceleration, and b and a the velocity and acceleration of the cylinder. The subscript "n" denotes the component of the related vector in the direction normal to the axis of the cylinder. Water particle velocity and acceleration are the superposition of those of currents and waves. In the presence of ocean currents, wave frequencies may be shifted due to the Doppler effect, which is neglected in our computation because it is assumed that current velocity is small in comparison with the phase velocity of incident waves.

Forces applied on the truncated bottom of a cylinder in the axial direction include the integration of wave pressure over the bottom S_B, and drag and added-mass forces, which are equivalent to half of a thin circular disk of the same diameter of the cylinder, in heave motion (Sarpkaya and Isaacson 1981).

$$F_t = \rho_f \iint_{S_B} \left(\frac{\partial(\phi^{(1)} + \phi^{(2)})}{\partial t} + \frac{1}{2} \left| \nabla \phi^{(1)} \right|^2 \right) ds + C_{mt} \rho_f \frac{4}{3} \left(\frac{D}{2} \right)^3 \left[(a_f)_t - a_t \right] + \frac{1}{2} \rho_f C_{Dt} \frac{\pi D^2}{4} \left| (v_f)_t - v_t \right| \left[(v_f)_t - v_t \right] \tag{5-2}$$

where $\phi^{(1)}$ and $\phi^{(2)}$ are first- and second-order potential of incident waves, and C_{mt} and C_{Dt} are the added-mass and drag coefficient of the truncated cylinder bottom, respectively. $(v_f)_t - v_t$ and $(a_f)_t - a_t$ are the relative velocity and acceleration of the cylinder bottom to ambient fluid in the axial direction, respectively. Wave kinematics and first- and second-order incident wave potential used in the above equations are computed using a HWM (Zhang et al. 1996).

To account for vortex induced motion (VIM) of a spar in the presence of strong currents, such as loop currents in the Gulf of Mexico, an additional term representing the lifting force (or transverse force) applied on per unit length on the cylinder is added into the Morison equation.

$$dF_l = \frac{1}{2} \rho_f C_L D v_c^2 \cos(2\pi f \times t) \vec{e}_t \times \vec{e}_c \tag{5-3}$$

where $\vec{e}_c$ and $\vec{e}_t$ are the unit vectors in the current direction and the axial direction, respectively, C_L the lifting coefficient, and f the vortex shedding frequency. It is related to the Strouhal number (S_o), defined by

$$S_0 = \frac{fD}{v_n} \tag{5-4}$$

The Strouhal number and lifting coefficient in the context of a spar equipped with helical strakes on its surface and constrained by its mooring/riser systems are not well documented. In our computation, they were determined by fitting the mean and the average one-third and one-fifth amplitude and period of the simulated and measured LF sway of a spar model. In the case of a Deep-Star spar, the fitting yields $S_o = 0.25$ and $C_L = 0.45$. The procedure for determining the lifting coefficient is illustrated in Figure 5-1.

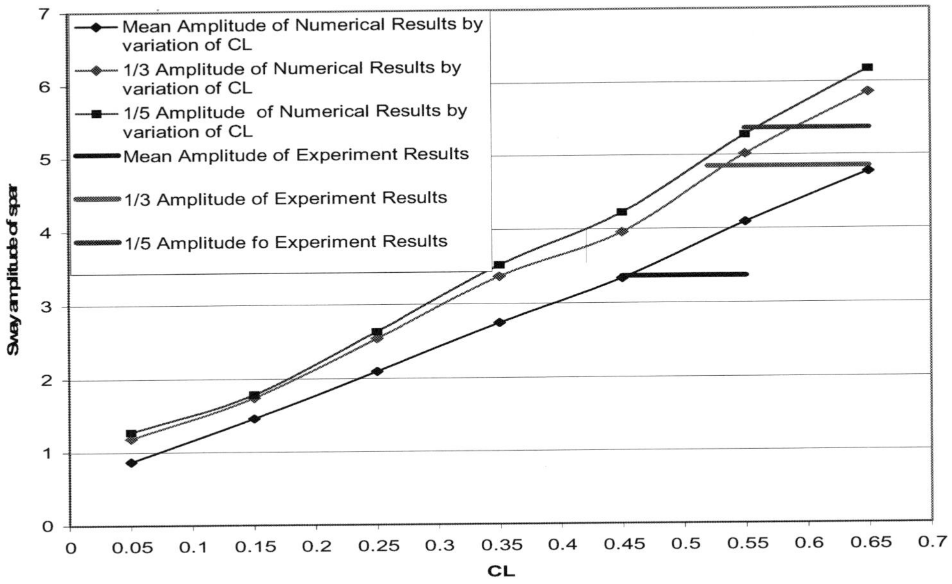

Fig. 5-1. The value of C_L vs. sway amplitude

Considering that the velocities of currents and waves may change along the longitudinal axis of a spar, the total wave and current loads on the spar are computed through the numerical integration of the corresponding loads over a number of segments along its axis.

The computation of wind force is based on the empirical formula recommended by API (RP-2A)

$$F_{wind} = \frac{1}{2} C_s \rho_a V_a^2 A \tag{5-5}$$

where ρ_a is the density of air, C_s the shape coefficient depending on the shape of superstructure of a spar and wind direction, V_a the total wind velocity, and A the projected area of a spar above the sea level.

Modeling Wave, Wind, and Current

As an input to COUPLE, incident ocean waves can be described by either time series of measured wave elevation or a wave spectrum. To compare simulated motion and forces of a floating structure with the corresponding measurements, measured wave elevations are often used as the input. The amplitude and initial phase of free waves in a measured irregular long-crested wave train are calculated as a function of frequency using the decomposition part of an HWM. The amplitude and initial phase of free waves are then used in the prediction part of the same HWM for computing wave kinematics along the longitudinal axis of a spar as a function of time (Zhang et. al. 1996). In the absence of measured waves, a typical wave spectrum, such as a JONSWAP spectrum, is used as the input. The related free (linear) wave spectrum is obtained by matching its resultant spectrum (including bound waves) with a given analytic wave spectrum selected for the simulation. The amplitude of free-wave components and their initial phase are obtained using a random phase method or random Fourier coefficient method (Tuah and Hudspeth 1982) based on a free-wave amplitude or energy spectrum.

The gustiness of wind is simulated based on a spectral energy density recommended by API rules (RP-2A).

$$\frac{fS(f)}{\sigma(z)^2} = \frac{f / f_p}{(1 + 1.5 f / f_p)^{5/3}} \tag{5-6}$$

where $S(f)$ is the spectral energy density, z_a the vertical coordinate of the wind pressure center of a floating structure, f the frequency, $\sigma(z)$ the standard deviation of wind speed, and f_p is given by,

$$f_p z_a / \overline{V}_a = 0.025 \tag{5-7}$$

$\overline{V}_a$ is the average wind velocity in a 1-hour period. Based on a wind spectral density function, the total wind velocity is calculated following the procedures similar to those for simulating random water waves.

In the current version of COUPLE, both velocity and direction of currents are the input, which is a function of the vertical coordinate *(z)* but assumed to be steady. However, the extension to allow for unsteady currents can be made without principle difficulties.

Dynamic Equations for the Hull

The equations of linear motion of a rigid body expressed in the $\hat{o}\hat{x}\hat{y}\hat{z}$ (fixed in space) coordinates and its rotational motion in the *oxyz* (fixed on the hull) coordinates are:

$$m\frac{d^2\xi}{dt^2} + mT^t(\frac{d\omega}{dt} \times r_g) + mT^t(\omega \times (\omega \times r_g)) = F_t \tag{5-8}$$

$$I_o\frac{d\omega}{dt} + \omega \times I_o\omega + mr_g \times (T\frac{d^2\xi}{dt^2}) = M_o \tag{5-9}$$

where $\xi = (\xi_1, \xi_2, \xi_3)^t$ is the coordinates of the origin of the *oxyz* in the $\hat{o}\hat{x}\hat{y}\hat{z}$ coordinates and $\frac{d^2\xi}{dt^2}$ is its acceleration. $\omega = (\omega_1, \omega_2, \omega_3)^t$ is the angular velocity of the hull, $r_g = (x_g, y_g, z_g)^t$, the vector from the origin of the *oxyz* coordinates to the gravitational center of the hull, and I_o the moment of inertia of the hull defined in the *oxyz* coordinates. F_t is the total force applied on the hull and expressed in the $\hat{o}\hat{x}\hat{y}\hat{z}$ coordinates, which includes wave, wind, current loads, hydrostatic restoring force, and restraining force from mooring lines and risers. M_o is the total moment and is expressed in the *oxyz* coordinates. T is a transfer matrix between the hull-fixed coordinates (*oxyz*) and the space-fixed coordinates ($\hat{o}\hat{x}\hat{y}\hat{z}$). It is an orthogonal matrix with the property that $T^t = T^{-1}$.

Equations for a Flexible Slender Rod

Our computation of the motion and tension of a flexible slender rod (mooring lines or risers) mainly follows Garrett (1982). To allow for large extension elements, such as springs or polyester ropes, Chen et al. (2001) extended his formulation, which is described below. The instantaneous configuration of a rod is denoted by a vector, $\mathbf{r}(s,t)$, a function of time, and the arc length along the rod. The dynamic and constrain equations of a rod of infinitesimal extension were given by

$$-(Br'')'' + (\lambda r')' + q = \rho \ddot{r}, \quad \lambda = T - Bk^2 \tag{5-10}$$

$$r' \times r' = (1+\varepsilon)^2 \tag{5-11}$$

where B is the bending stiffness, T the tension, and k the curvature of the rod. q is the external force applied on the rod per unit length, and ρ, the mass per unit length. The prime and over dots stand for the partial derivatives with respect to s and time, respectively. $\mathcal{E}$ denotes the strain of the rod, which is equal to T/EA where E is the Young's modulus and A, the cross section area of the rod. It is noted that the strain of the rod is explicitly considered in Eq. 5-11 but neglected in Eq. 5-10. For the elements in a mooring line with a large extension, such as a spring or polyester rope, their bending moment is small and can be neglected for simplicity. Dropping the term involving the bending moment, the equation corresponding to Eq. 5-10 is derived, which considers relative large elongation ($\mathcal{E}$).

$$\left(\frac{\lambda}{1+\varepsilon} r'\right)' + q(1+\varepsilon) = \rho \ddot{r} \tag{5-12}$$

In using a finite element method to solve the preceding equations, the shape and tension of each element are approximated by a cubic and quadratic spline function, respectively. A Newton method is used to solve static equations, and a Newmark-β method is used to solve dynamic equations in the time domain.

Because the modulus of a polyester rope depends on the tension, the empirical formula given by Del Vecchio (1992) is employed.

$$E = \alpha + \beta L_m - \gamma L_a - \delta Log(T) \tag{5-13}$$

where α, β, γ, and δ are constants, related to the main characteristics of a polyester rope, L_m is the mean tension, and L_a and T are the amplitude and period of dynamic tension, respectively. In COUPLE, δ is set to zero because the dependence of the modulus on the period of a dynamic load is insignificant (Bosman and Hooker 1999). Even at the mean position of a spar experiencing wind, current, and wave loads, the modulus of polyester ropes in different mooring lines of an integrated mooring system is different because of different mean tensions. To determine the modulus of each mooring line, we first let $L_a = 0$ and calculate the modulus and tension of each polyester rope through iteration. Our calculation indicates that the mean (static) tension and modulus of each mooring line converge rapidly just after two or three iterations. Based on the updated modulus of each rope, the simulation of the motion of a moored spar and the tension in mooring lines is made, given the met-ocean conditions. Since the dominant responses of a spar are its LF motions, the amplitude of dynamic tension in a polyester mooring line is also dominated by the corresponding LF tension. Using a low-pass filter, the average amplitude of dynamic tension in a polyester rope can be determined as sketched in Figure 5-2. Knowing the approximate L_a , the modulus of each polyester line is updated, and the dynamic simulation of a spar positioned by an integrated polyester mooring system is repeated. The iteration terminates if the relative difference in the modulus of two consecutive iterations is smaller than a prescribed error tolerance. More detailed description of the iterative procedure for determining the modulus of polyester mooring lines is given by Kim et. al. (2003).

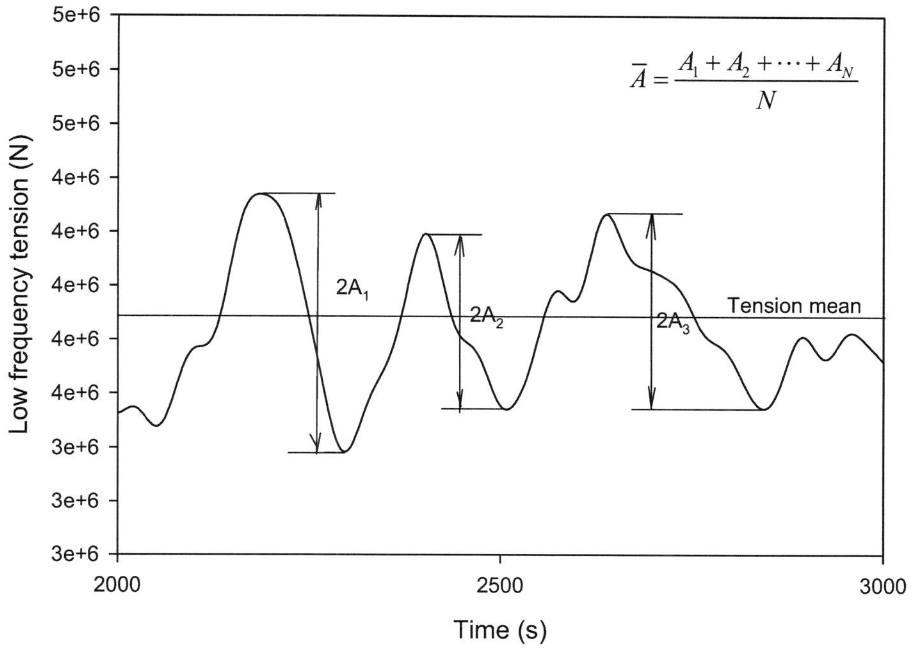

Fig. 5-2. Sketch for computing the average amplitude of dynamic tension

Coupling Between a Hull and Its Mooring/Riser Systems

Motion equations for a hull and its mooring and riser systems are coupled by imposing prescribed conditions at their connections (fairleads or porches). For example, if a hinge connection is imposed between the hull and its mooring lines, then the forces and displacements of the hull and a mooring line at its fairlead are the same, and no moment is applied there. More complicated connection conditions can be simulated by appropriately imposing force, moment, and relative displacement on the hull and its mooring/riser systems at their connections. The coupled equations for the hull and its mooring and riser systems are then solved simultaneously in the time domain using a Newmark-β method. At each time step, the velocities and positions of the hull and all mooring lines/risers are first predicted based on the velocities, positions, and accelerations at previous step. Then the correctors for positions, velocities, and accelerations are calculated based on the dynamic equations. If the difference between the two correctors of consecutive iterations is less than a prescribed error tolerance, the simulation moves forward to the next step.

Couple Analysis for a Classic Spar

COUPLE was used to simulate the global motions of a moored floating structure, such as spars and a mini tension leg platform (TLP). The numerical results were examined against the corresponding laboratory measurements and satisfactory consistency between them was observed (Chen et al. 2000, 2002). Described below is an example of simulating a spar positioned by a mooring system consisting of 14 integrated polyester mooring lines and deployed in water depth of 10,000 ft. The characteristics of the spar studied here are similar to those of the Deep-Star spar, a classic spar. The main characteristics of the hull and its mooring system are given in Tables 5-1 and 5-2, and the configuration of the mooring system is sketched in Figure 5-3. The static offset curve of the mooring system is plotted in Figure 5-4. To demonstrate the necessity of using the extended CABLE3D based on the large elongation formulation, also plotted in the figure is the corresponding curve computed using a similar code but based on the small elongation assumption. Near the mean position (~ 17.5m) of the spar experiencing a 100-year hurricane in the Gulf of Mexico, the figure shows that the restraining force of the mooring system is reduced by 16 percent when the elongation of polyester ropes is considered.

Table 5.1. The Main Specification of the Spar

Hull	
Displacement	53,600 m.ton
Total displacement	220,640 m.ton
Diameter	122 ft.
Length	705 ft.
Draft	650 ft.
Hard tank depth	220 ft.
KB	540 ft.
KG	462 ft.
KG (based on total displacement)	314 ft.
Radius of gyration	Pitch=221ft., yaw=28.5 ft.
Drag force coefficient	1.16
Wind force coefficient	0.0558 (kips/(ft./sec)^2)
Center of pressure	722 ft. ABL

Table 5.2. Mooring System Specification

MOORING LINE	
Mooring type	Poly taut
Mooring pattern	14-point, taut-leg omni-directional spread
Mooring line composition	Platform section 300 ft. × 5-1/8 in. K4 studless chain
	Middle section 13,300 ft. × 9.06 in. polyester
	Ground section 400 ft. × 5-1/8 in. K4 studless chain
Fairlead location	300 ft. ABL
Pretension	650 kips

Table 5.3. Hydrodynamic Force Coefficients

	Normal drag coefficient	Added-mass coefficient	VIV lifting force coefficient	Strouhal number
Spar	1.16	1.00	0.45	0.2
Chain	2.45	1.40	N/A	N/A
Rope	1.20	1.00	N/A	N/A

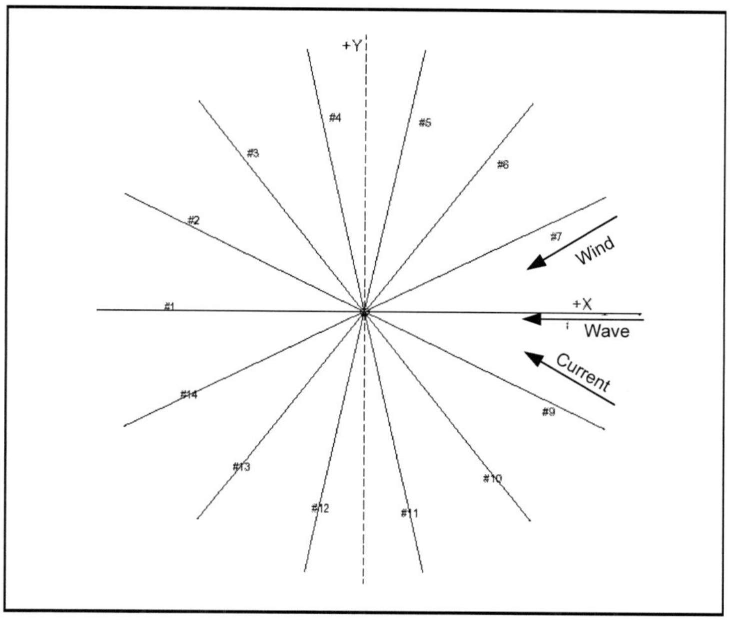

Fig. 5-3. A spread mooring system

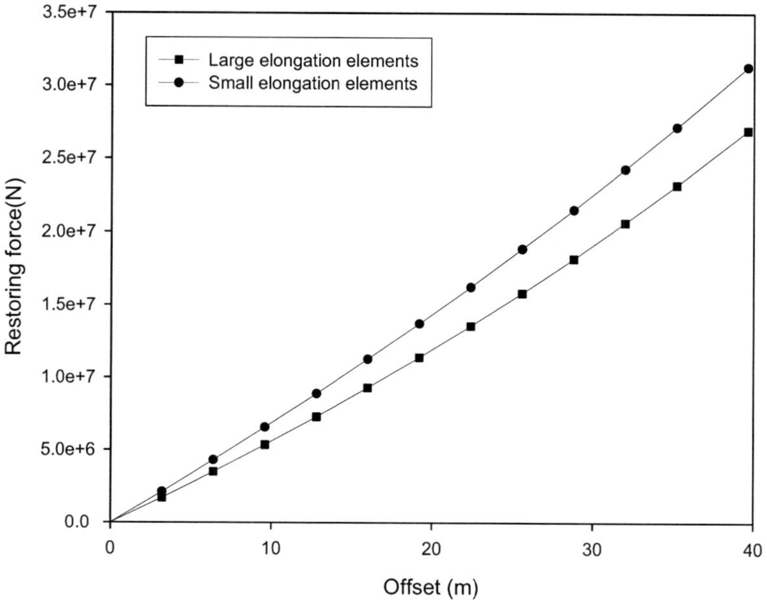

Fig. 5-4. Static offset curves based on large or small elongation codes

The met-ocean conditions used in our numerical simulation were intended to model a 100-year hurricane in the Gulf of Mexico. Long-crested incident waves were generated according to a JONSWAP spectrum of a significant wave height 41ft., a peak period 12.64 s, and a sharp factor 3.3. The steady wind speed was 98.4 mph. The velocity of the current at surface was 1.2 m/s and diminished with water depth as depicted in Figure 5-5. The directions of waves, wind, and current are specified in Figure 5-3.

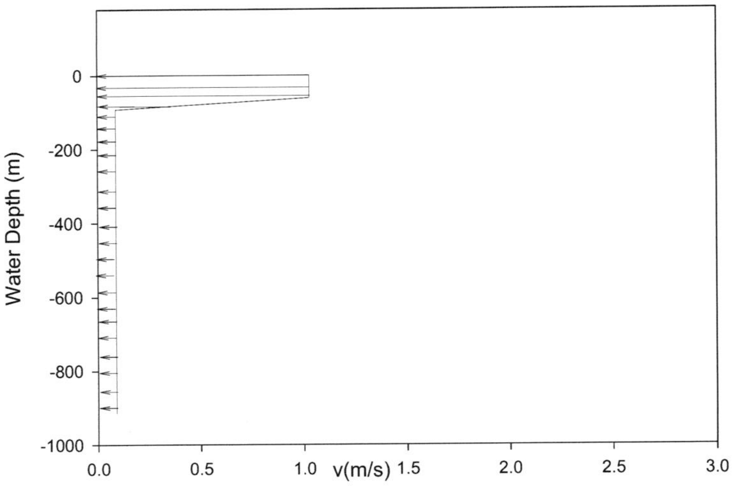

Fig. 5-5. The profile of current velocity

The added-mass, drag, and lifting coefficients of the hull and the Strouhal number are given summarized in Table 5-3, which were used in the Morison equation for computing wave and current loads on the hull. Also included in the table are the added-mass and drag coefficients of steel chains and a polyester rope consisting of an integrated mooring line. Previously, COUPLE was used to compute the same spar but positioned by a steel mooring system and deployed in 3000-ft. water (Ding et al. 2003). The comparison with the corresponding measurements of the model tests indicated the simulation using COUPLE was satisfactory. The differences between the present and previous simulation are the water depth and the composition of mooring lines. Three-hour numerical simulation was conducted using COUPLE. Typical surge, sway, and heave of the hull under the impact of a 100-year hurricane in the Gulf of Mexico are shown as a function of time in Figures 6-8, and the related amplitude spectra in Figures 5-9 through 5-11. As shown in these figures, the surge and sway are dominated by the slow-drift motion. Typical amplitudes of the slow-drift surge range from 4 m to 8.5 m and those of sway from 2 m to 4 m. The amplitudes of the heave range from 1 m to 2 m. The average periods of the slow-drift surge and sway are similar, about 190 s, and that of heave is about 30 s, which is close to the corresponding natural periods determined by numerical simulation of free decay tests of the spar. As expected, the frequencies of slow-drift surge and sway and that of heave are well below the WF.

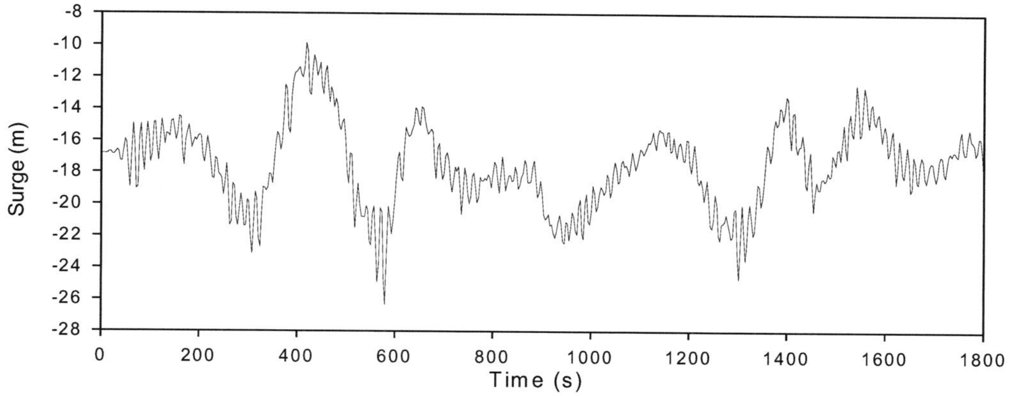

Fig. 5-6. Surge of the hull as a function of time

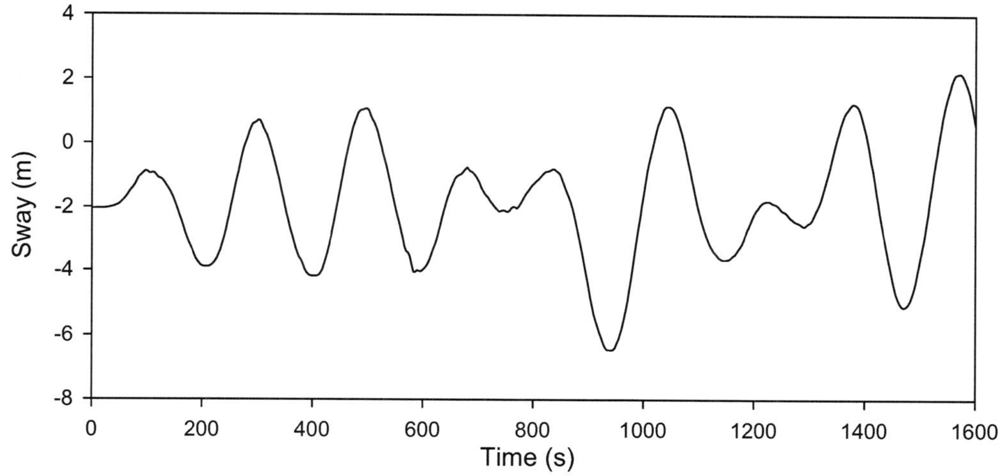

Fig. 5-7. Sway as a function of time

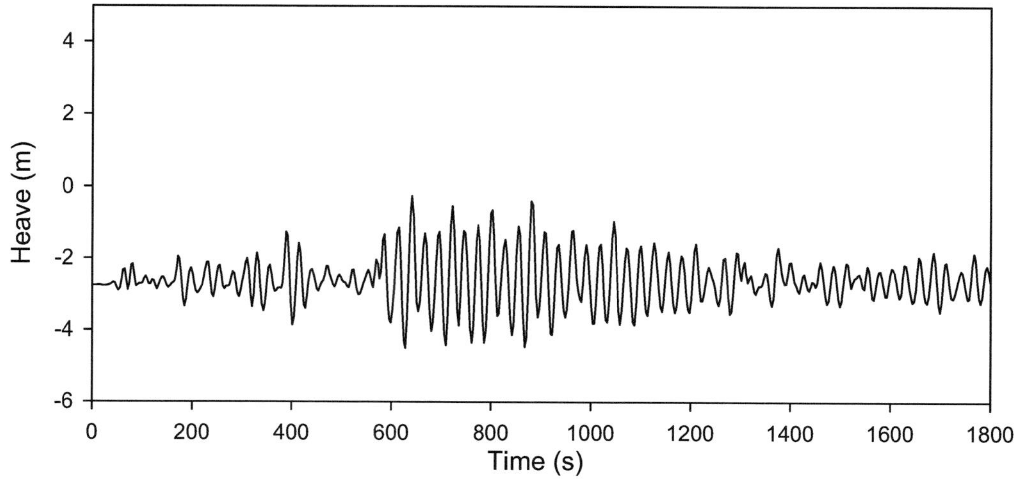

Fig. 5-8. Heave of the hull as a function of time

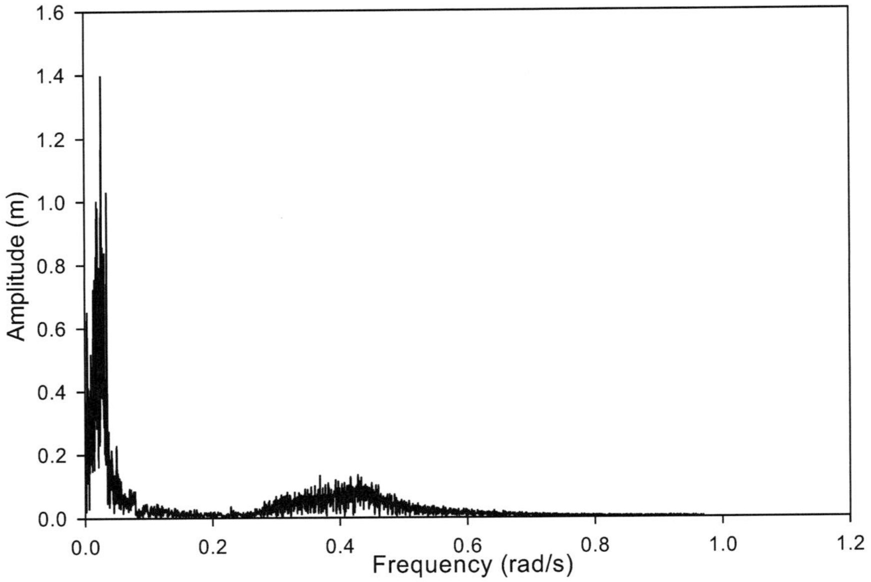

Fig. 5-9. Surge amplitude spectrum

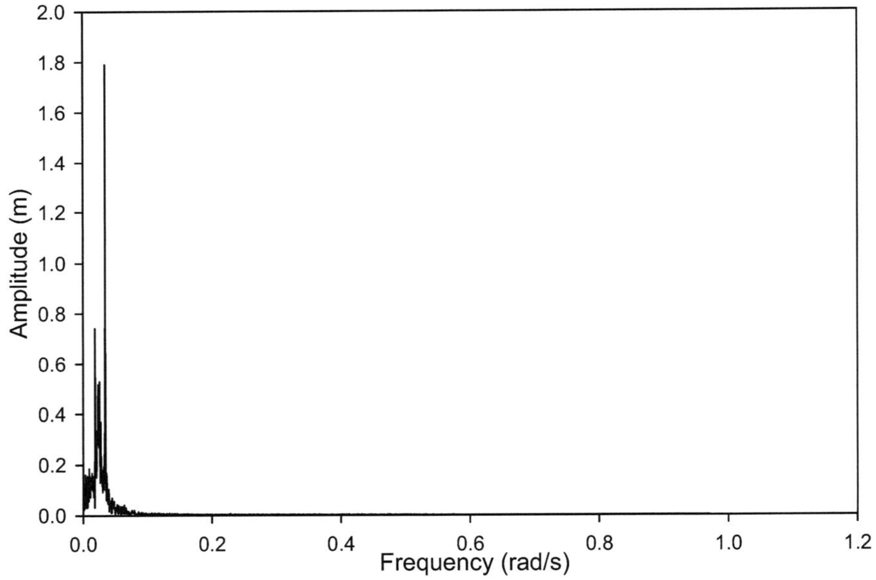

Fig. 5-10. Sway amplitude spectrum

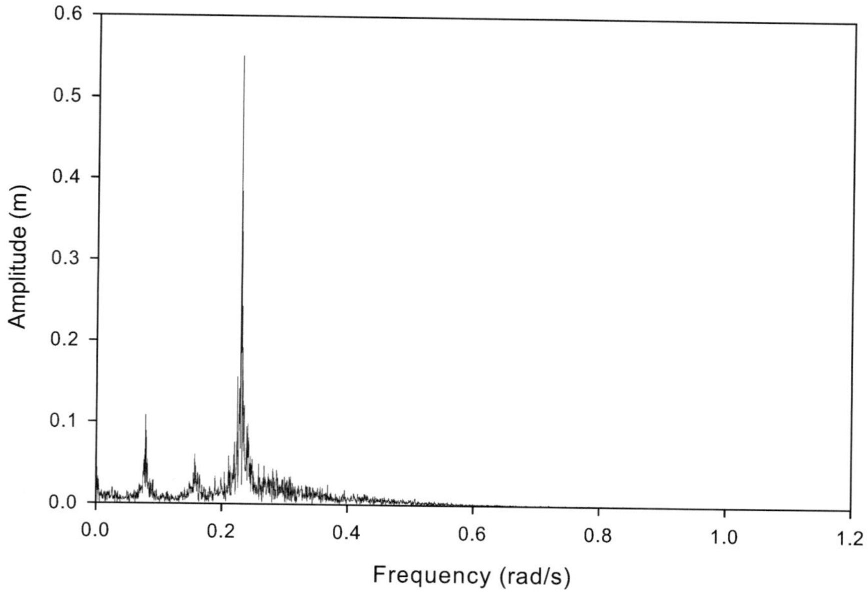

Fig. 5-11. Heave amplitude spectrum

The related tensions in mooring lines, #8 (windward side) and #1 (leeward side) are given as a function of time in Figures 5-12 and 5-13 and the related tension amplitude spectra in Figures 5-14 and 5-15. The maximum tension in both lines seems dominated by the slow-drift surge and sway of the hull. However, the tension caused by the heave of the hull is also significant.

To provide an overall picture of a 3-hour simulation, we summarize the statistics of the global motions of the hull in Table 5.4 and those of tensions in all 14 mooring lines in Table 5.5.

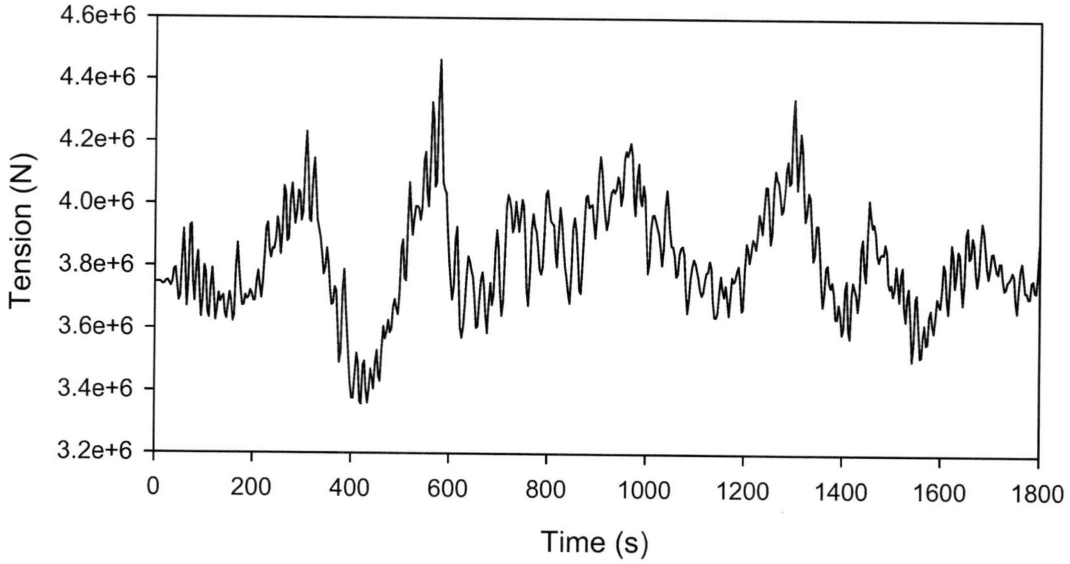

Fig. 5-12. Tension time series in line #8

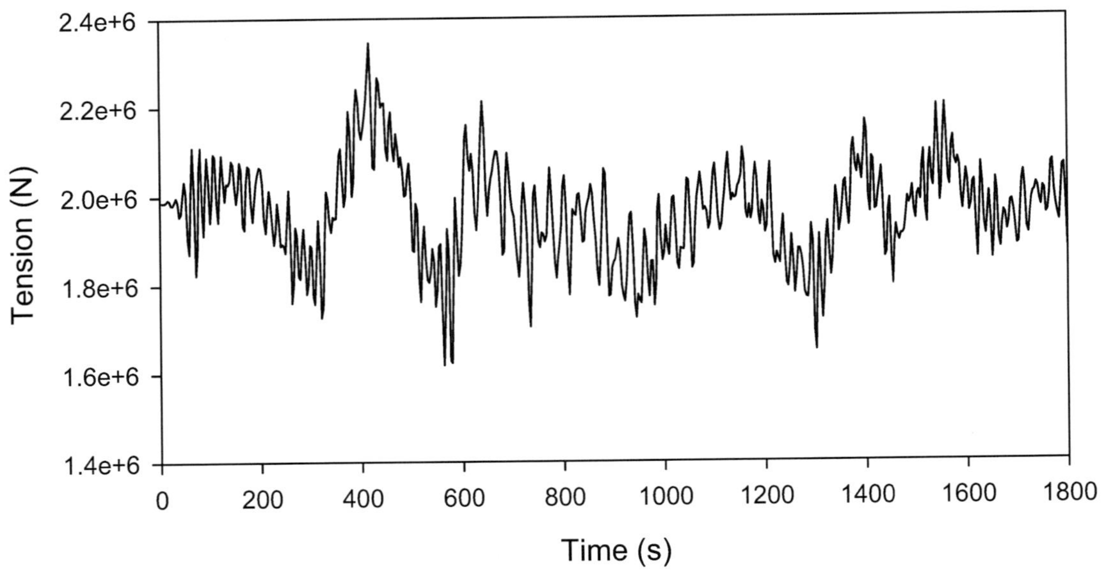

Fig. 5-13. Tension time series in line #1

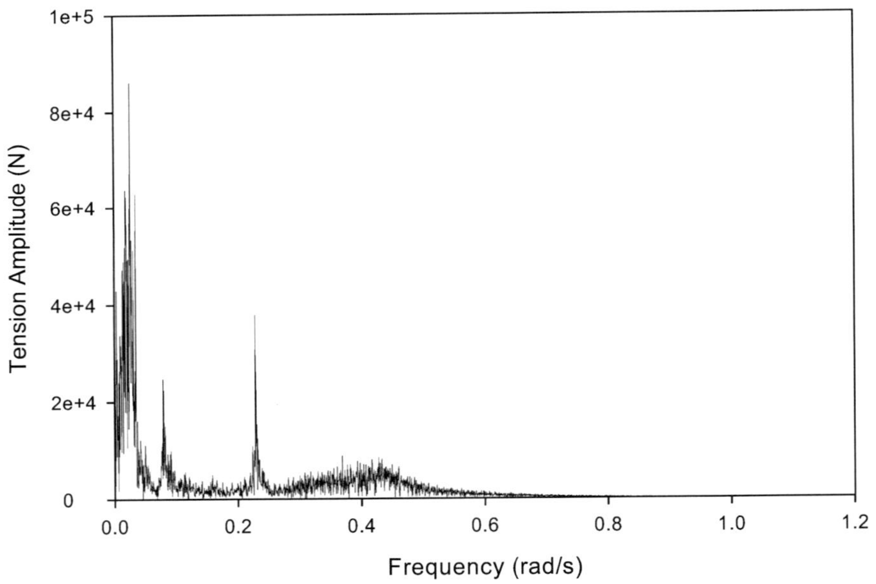

Fig. 5-14. Line # 8 (windward side) tension amplitude spectrum

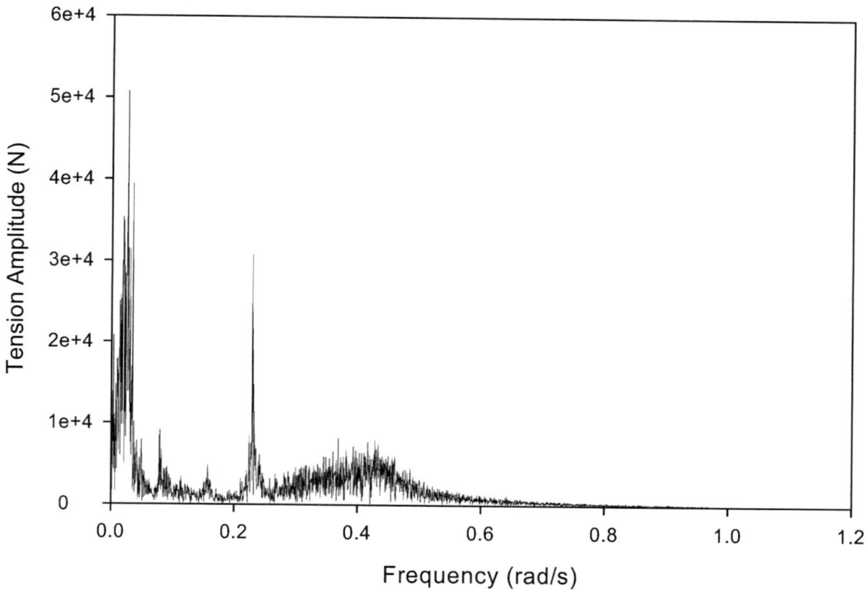

Fig. 5-15. Line # 1 (leeward side) tension amplitude spectrum

Table 5-4. Statistics of the Global Motions of the Spar

	Mean	Standard Deviation	Max
Surge (m)	-17.646	3.236	33.9
Sway (m)	-1.884	1.810	6.462
Heave (m)	-2.585	0.719	4.617
Roll (degree)	0.736	0.805	2.932
Pitch (degree)	-2.175	2.230	11.765
Yaw (degree)	-0.074	0.196	0.856

Table 5-5. Statistics of Tension

	Mean (N)	Standard Deviation (N)	Maximum (N)
Line # 1	1.961E+06	1.302E+05	2.516E+06
Line # 2	1.993E+06	1.347E+05	2.549E+06
Line # 3	2.155E+06	1.300E+05	2.638E+06
Line # 4	2.458E+06	1.139E+05	2.820E+06
Line # 5	2.874E+06	1.020E+05	3.169E+06
Line # 6	3.315E+06	1.296E+05	3.878E+06
Line # 7	3.664E+06	1.784E+05	4.509E+06
Line # 8	3.832E+06	2.121E+05	4.830E+06
Line # 9	3.775E+06	2.141E+05	4.745E+06
Line # 10	3.509E+06	1.830E+05	4.245E+06
Line # 11	3.100E+06	1.323E+05	3.507E+06
Line # 12	2.657E+06	9.592E+04	2.977E+06
Line # 13	2.289E+06	1.008E+05	2.657E+06
Line # 14	2.056E+06	1.184E+05	2.540E+06

Acknowledgment

The authors would like to thank Offshore Technology Research Center (OTRC) of Texas A&M University for the financial support of this work.

References

API RP-2A, (1997). Recommended Practice for Planning, Designing and Constructing Fixed Offshore Platform. American Petroleum Institute, Washington, DC, USA.

Bosman, R. L. M., and Hooker, J. (1999). "Elastic Modulus Characteristics of Polyester Mooring Ropes." *Proceedings of the Annual Offshore Technology Conference*, v3, 139-143

Cao, P.M. and Zhang, J. (1997). "Slow motion responses of compliant offshore structures." *International Journal of Offshore and Polar Engineering* 7(2), 119-126.

Chen, X.H., Zhang, J. and Ma, W. (1999). "Coupled Analysis of a JIP Spar and Its Mooring System." *Proceedings of the International Offshore and Polar Engineering Conference* (ISOPE) Vol. I, 293-300.

Chen, X.H., Zhang, J., Johnson, P. and Irani, M. (2000). "Studies on the dynamics of truncated mooring line." *Proceedings of the International Offshore and Polar Engineering Conference* (ISOPE), Vol.2, pp. 94-101.

Chen, X.H., Zhang, J, and Ma, W. (2001). "On Dynamic Coupling Effects Between a Spar and Its Mooring Lines." *Ocean Engineering* 28 (2001) pp. 863-887.

Chen, X., Zhang, J., Liagre, P., Niedzwecki, J. and Teigen. P. (2002). "Coupled Dynamic Analysis of a Mini TLP: Comparison with Measurements." 21st International Conference on Offshore Mechanics and arctic Engineering, paper Number OMAE 02-8536.

Del Vecchio, C.J.M. (1992). "Light Weight Material for Deep Water Mooring." Ph D. Thesis, University of Reading, UK.

Ding, Y., Kim, M., Chen, X. and J. Zhang (2003) "A Numerical Code (COUPLE6D) for Coupled Dynamic Analysis of Moored Offshore Structures." *Proceedings of International Symposium on Deep-Water Mooring System*, p168-182.

Garrett, D.L., (1982). "Dynamic Analysis of Slender Rods." Journal of Energy Resources Technology, Transaction of ASME 104, 302-307.

Lee, C.H. (1995). *WAMIT Theory Manual*. Report No. 95-2, Massachusetts Institute of Technology, Cambridge, MA, USA.

Kim, M., Ding, Y. and Zhang, J. (2003). "Dynamic Simulation of Polyester Mooring Lines," *Proceedings of International Symposium on Deep-Water Mooring Systems*, p101-114.

Ma, W. and Webster, W.C. (1994). *An Analytical Approach to Cable Dynamics: Theory and User Manual*. Sea Grant Project R/OE-26.

Sarpkaya, T. and Isaacson, M., (1981). *Mechanics of Wave Forces on Offshore Structures*. Van Nostrand Reinhold Company Inc., New York, NY, USA.

Tuah, H. and Hudspeth, R.T. (1982). "Comparison of Numerical Random Sea Simulations," *Journal of Waterway, Port, Coastal and Ocean Divisions*, ASCE, 108, (WW4), Proceedings Paper 17488, pp. 569-584.

Zhang, J., Chen, L., Ye, M. and Randall, R.E. (1996). "Hybrid Wave Model for Unidirectional Irregular Waves. Part I. Theory and Numerical Scheme." *Applied Ocean Research*, Vol. 18, 77-92.

Zhang, J., Yang, J., Prislin, I., Wen, J., and Hong, K. (1999). "Deterministic Wave Model for Short Crested Ocean Waves, Part I. Theory and Numerical Scheme." *Applied Ocean Research*, Vol. 21, 167-188.

Chapter 6: Spar Analysis, Comparison, and Theory: Morison Formula versus Diffraction Theory

By Iftekhar Anam, Associate Professor, Department of Civil Engineering, The University of Asia Pacific; Dhaka, Bangladesh; and José M. Roësset, Professor, Department of Civil Engineering, Texas A&M University, and Hon.M.ASCE

Abstract

The second-order difference-frequency forces on spar platforms can be evaluated analytically and numerically using various nonlinear hydrodynamic force models. The models studied and compared in this chapter are the full time-domain Morison's equation (ME), a second-order Morison's equation (ME2), the second-order diffraction-radiation theory (Φ2 theory), a Φ2 theory assuming very slender structural dimensions (slender Φ2 theory) and Newman's approximation. The purpose is to show the effect of different nonlinear forces on the dynamic response of spars and to illustrate their basic differences. Analytical and numerical results show the limitations of Morison's equation as the slender-body counterpart of the diffraction theory for nonlinear problems. A new combined-force method is also suggested to approximate the second-order difference-frequency forces from the Φ2 theory with considerably less computational effort. The new method is formulated by combining two limiting cases of the Φ2 theory—Newman's approximation and the "slender Φ2 theory." Numerical results show that the new method reproduces the individual nonlinear effects of the Φ2 theory better than the other approximations.

Nonlinear Forces on Spars

Flexible offshore structures like spars are designed to have natural periods of vibration much longer than the dominant wave periods so that there are hardly any linear (first-order) forces at the natural frequencies. The response to the wave-excitation force is largely inertial and is often smaller than the ones induced by the nonlinear interaction of waves. Moreover, the linear response can be well predicted by linear wave theory (LWT) or any approximate nonlinear extension of it. Thus, much of the work on the dynamic response of spars has been focused on numerically estimating the nonlinear forces and responses accurately and comparing them to the available lab results (for example, Mekha et al. 1995, 1996; Ran et al. 1996, 1997, 1999; Weggel & Roësset 1996a, 1996b; Cao & Zhang 1996; Jha et al. 1997, and others).

For slender bodies with a D/L (diameter/wavelength) ratio less than 0.20, Morison's equation (Morison et al. 1950) or its extension by Rainey (1989) is often considered to be adequate in calculating hydrodynamic forces. Diffraction effects are considered insignificant for slender bodies and are neglected in these formulations. In a time domain solution using Morison's equation (ME), the higher order nonlinearities are implicitly included. A solution in the frequency domain, on the other hand, requires explicit calculation of wave-forces up to a desired order using perturbation theory.

The other approach to wave-force calculation is based on diffraction theory (MacCamy and Fuchs 1954). Here the fluid is considered inviscid, which excludes any drag term from the force equation. However, it accounts for diffraction (due to the width of the body) and radiation effects (caused by the structural motions). Diffraction theory is also developed using perturbation techniques, and therefore, the forces can only be calculated up to a certain order whether in the time domain or the frequency domain.

This chapter is mainly concerned with second-order, difference-frequency effects. Several studies have indicated that second order (difference-frequency) is an adequate truncation point for spars. Additionally for H/D (wave-height/diameter) ratios typical of spars, the inertia force is by far the dominant force. For the inertia force with both slender-body approximation and diffraction theory, the second-order effects discussed in this chapter include the difference-frequency forces caused by

1. the temporal acceleration due to second-order potential;
2. the convective acceleration;
3. the axial divergence;
4. the free-surface fluctuation; and
5. the structural displacements (both translation and rotation).

In addition, the nonlinear drag force in Morison's equation, even when statistically linearized, causes two types of nonlinearities due to

1. structural displacements, that is, calculation of linearized drag force at the displaced position of the structure; and
2. free-surface fluctuation, that is, integration of the drag force up to the free surface.

The drag also causes a mean force on the structure, which can be significant particularly in the presence of current.

These effects are evaluated first for Morison's equation, and their counterparts are sought in the diffraction theory formulation. Similar work was carried out numerically by Kim and Chen (1994) for an ALP (Articulated Leg Platform). Here the comparisons are made analytically, and their basic differences are pointed out. In this context, the approximation suggested by Newman (1974) is also discussed.

This chapter attempts to simulate all the second-order, difference-frequency effects present (directly or indirectly) in the time domain approach. The calculations involve some elementary trigonometric manipulations to get difference-frequency forces for bichromatic waves. The total effects for irregular waves are summations for several such combinations.

All the formulations discussed involve unidirectional waves (in the x-direction) and spar displacements along the x and z axes with rotation about the y-axis.

Calculation of Wave Forces (Morison's Equation)

The Force Equation

For deep-draft slender structures like spars, typical values of the diffraction parameter kR (near the dominant wave frequency) are small (about 0.20 to 0.40). As such, the use of Morison's equation is often considered as valid as the more rigorous diffraction theory and has been used in several studies. This is the first approach followed in this chapter.

Only the second-order, difference-frequency effects in the horizontal exciting force are derived for mathematical convenience for use in the problem-formulation and comparison with the second-order diffraction theory (Φ2 theory).

According to the (modified) Morison's equation, the horizontal wave-force and moment on a differential vertical segment (dz) are given (after adjustments for structural acceleration) by

$$dF_x = [K_I a_x + K_D|u_r|u_r + K_m u_r \partial w/\partial z]\, dz \;;\; \text{and } dM_{y0} = z\, dF_x \qquad (6\text{-}1)$$

where $K_I = \rho C_I A$, $K_D = \rho C_D R$, $K_m = \rho C_m A$

ρ = water density, C_I = inertia coefficient, C_D = drag coefficient, $C_m = C_I - 1$,
A = cross-sectional area, R = radius = half-width of projected surface,
a_x = horizontal acceleration = $du/dt = \partial u/\partial t + u\, \partial u/\partial x + w\, \partial u/\partial z$,
u_r = relative horizontal velocity = horizontal (wave – structural) velocity.

In Eq. 6-1, the first term gives the inertia force, the second is the drag, and the third term corresponds to the axial divergence.

The total horizontal force is $F_x = \int dF_x$ and the moment about the spar-CG is $M_{cg} = \int(z - z_{cg})dF_x$, where $\int$ implies integration between z = –L (the spar-bottom) and z = η (the instantaneous wave-elevation), and z_{cg} is the z-coordinate of the spar's CG (negative and close to –L/2, as the CG is below the MWL).

For a spar, the inertia force is normally the primary excitation because the Keulegan-Carpenter (K-C) number is typically very small, that is, wave amplitude << RC_I/C_D. For example, in the sample case used in this chapter, the spar is subjected to maximum wave amplitudes of around 3 m compared to $RC_I/C_D \cong 130$ m. Drag effects contribute to the hydrodynamic damping and control the magnitude of the nonlinear peaks in the spectral response at the natural frequencies, but the total excitation force is controlled by the inertia.

Second-Order Morison's Equation (ME2)

As an alternative to the complete Morison's equation in time domain (t-d), its second-order approximation can also be used. Only the difference-frequency forces are important for flexible structures because their natural frequencies are smaller than the typical range of wave frequencies. For example, the surge and pitch natural frequencies of the spar studied later in this chapter are only 0.019 and 0.095 rad/sec., respectively, whereas the dominant wave frequencies are around 0.40~0.50 rad/sec.

Up to second order, the force-equation is

$$F_x = \int^{(\eta)} K_I\, \partial u^{(1)}/\partial t\, dz + \int K_I(\partial u^{(2)}/\partial t + u^{(1)}\partial u^{(1)}/\partial x + w^{(1)}\partial u^{(1)}/\partial z + \partial u^{(\delta)(1)}/\partial t)\, dz +$$
$$\int K_D|u_r^{(1)}|u_r^{(1)}\, dz + \int K_m u_r^{(1)}\partial w^{(1)}/\partial z\, dz \qquad (6\text{-}2)$$

where $\int^{(\eta)}$ is integration between 0 and $\eta^{(1)}$, $\int$ between $-L$ and 0, and $u^{(\delta)}$ is the horizontal velocity at the instantaneous position of the structure.

The forces in Eq. 6-2 are the following:

$\int^{(\eta)} K_I\, \partial u^{(1)}/\partial t\, dz$ is the combination of first-order and second-order inertia forces calculated by integrating the first-order inertia force up to the free surface level. Here, only the integration from the mean water level (MWL) to the free surface level is second order.

$\int K_I\, \partial u^{(2)}/\partial t\, dz$ is the second-order inertia force calculated by integrating the second-order wave acceleration up to the MWL.

$\int K_I\, (u^{(1)}\partial u^{(1)}/\partial x + w^{(1)}\partial u^{(1)}/\partial z)\, dz$ is the force due to the second-order convective acceleration (quadratic velocity) term.

$\int K_I\, \partial u^{(\delta)(1)}/\partial t\, dz$ is the second-order inertia force (calculated up to the MWL) due to the first-order structural displacements.

$\int K_D |u_r^{(1)}| u_r^{(1)}\, dz$ is the second-order drag term integrated up to MWL.

$\int K_m u_r^{(1)} \partial w^{(1)}/\partial z\, dz$ is the so-called second-order axial divergence term.

Assuming constant K_I, K_D, and K_m (that is, same coefficients and cross-sectional properties at any depth), these forces can be calculated analytically (Kim and Chen 1994, Anam 2000). The inertia force components derived from ME2 are listed in Tables 6-1 and 6-2 (for deep water).

Table 6-1. Second-Order Difference-Frequency Inertia Forces (Deep Water)

Type of Force	Resultant Force
First-order potential	$gK_I \Sigma a_i [r_i] \sin\theta_i$
Second-order potential	$gK_I \Sigma\Sigma a_* k_i (1-\omega_j/\omega_i)[r_-] \sin\theta_-$
Convective	$-K_I \Sigma\Sigma a_* \omega_* (k_-/k_+)[r_+] \sin\theta_-$
Axial Divergence (with displacement and rotation)	$K_m \Sigma\Sigma \omega_* \{a_*(r_+)\, k_-/k_+ + x_{Gjm}{}^{(1)} a_i{}^{(1)}(r_i) - x_{Gim}{}^{(1)} a_j{}^{(1)}(r_j) + \gamma_{jm}{}^{(1)} a_i{}^{(1)}[L/2(R_i)-(r_i)/k_i] - \gamma_{im}{}^{(1)} a_j{}^{(1)}[L/2(R_j)-(r_j)/k_j]\}(\sin\theta_-)/2$
Free-Surface Fluctuation	$gK_I \Sigma\Sigma a_* [k_i-k_j] (\sin\theta_-)/2$
Structural Surge	$gK_I \Sigma\Sigma \{x_{Gim}{}^{(1)} a_j{}^{(1)} k_j[r_j] - x_{Gjm}{}^{(1)} a_i{}^{(1)} k_i[r_i]\}(\sin\theta_-)/2$
Structural Rotation	$gK_I L\, \Sigma\Sigma \{\gamma_{im}{}^{(1)} a_j{}^{(1)} k_j[R_j] - \gamma_{jm}{}^{(1)} a_i{}^{(1)} k_i[R_i]\}(\sin\theta_-)/4$

Note: $r_i = 1-e^{-kiL}$, $R_i = 1+e^{-kiL}$, etc.

Table 6-2. Second-Order Difference-Frequency Inertia Moments (Deep Water)

Type of Moment	Moment about CG
First-order potential	$gK_I \Sigma a_i \{d_i\} \sin\theta_i$
Second-order potential	$gK_I \Sigma\Sigma a_* k_i (1-\omega_j/\omega_i) \{d_-\} \sin\theta_-$
Convective	$-K_I \Sigma\Sigma a_* \omega_* (k_-/k_+)\{d_+\} \sin\theta_-$
Axial Divergence (with displacement and rotation)	$K_m \Sigma\Sigma \omega_* \{a_*(d_+)k_-/k_+$ $+ x_{Gjm}{}^{(1)} a_i{}^{(1)}(d_i) - x_{Gim}{}^{(1)} a_j{}^{(1)}(d_j)$ $+ \gamma_{jm}{}^{(1)} a_i{}^{(1)}[(L^2/4+2/k_j^2)r_j - LR_j/k_j] - \gamma_{im}{}^{(1)} a_j{}^{(1)}[(L^2/4+2/k_i^2)r_i - LR_i/k_i]\}(\sin\theta_-)/2$
Free-surface fluctuation	$gLK_I \Sigma\Sigma a_* [k_i-k_j] (\sin\theta_-)/4$
Structural surge	$gK_I \Sigma\Sigma \{x_{Gim}{}^{(1)} a_j{}^{(1)} k_j d_j - x_{Gjm}{}^{(1)} a_i{}^{(1)} k_i d_i \}(\sin\theta_-)/2$
Structural rotation	$gK_I \Sigma\Sigma \{\gamma_{im}{}^{(1)} a_j{}^{(1)}[(k_j L^2/4+1/k_j)r_j - LR_j/2] -$ $\gamma_{jm}{}^{(1)} a_i{}^{(1)}[(k_i L^2/4+1/k_i)r_i - LR_i/2]\}(\sin\theta_-)/2$

Note: $d_i = L(1+e^{-kiL})/2-(1-e^{-kiL})/k_i$, etc.; assuming $z_{cg} = -L/2$

The main conclusions drawn from the analytical results on ME2 are the following:
The free surface fluctuation is the most important nonlinear effect. For surge, the convective term is almost as important and of a different sign.

The effects of displaced geometry and second-order potential are also important, and they work in the same direction as the free surface term. The effect of axial divergence is insignificant. The linear drag force is small compared to the inertia forces. But the difference-frequency drag forces contribute to the mean force, which is the only source of mean force in ME2. The mean force can be very important particularly in the presence of current. The drag force also contributes significantly to the system damping.

Wave Forces from Diffraction Theory

The Force Equation

In diffraction theory, the total velocity potential ϕ is taken as the summation of the incident wave potential ϕ_I, the diffraction potential ϕ_D (due to body-dimension), and the radiation potential ϕ_R (due to the structural motion). Using perturbation principles, the total potential can be written as the summation of first-order and higher-order terms.

$$\phi = \phi I + \phi D + \phi R$$

$$= \varepsilon\,(\phi I(1) + \phi D(1) + \phi R(1)) + \varepsilon 2\,(\phi I(2) + \phi D(2) + \phi R(2)) + \ldots\ldots\ldots \quad (6\text{-}3)$$

For regular shaped structures like cylindrical spars, the expressions for the first-order potentials $\phi_I^{(1)}$, $\phi_D^{(1)}$, $\phi_R^{(1)}$ and second-order difference-frequency potential $\phi_I^{-(2)}$ can be derived in explicit forms. Once the potentials are known, the hydrodynamic forces are computed directly by integrating the dynamic pressure over the body-surface, resulting in the following hydrodynamic force vectors

$$\text{First Order: } F^{(1)} = F_{I,D,R}^{(1)} + F_{HS}^{(1)} \tag{6-4}$$

$$\text{Second Order: } F^{(2)} = F_{I,D,R}^{(2)} + F_{HS}^{(2)} + F_q^{(2)} \tag{6-5}$$

Subscripts I, D, and R stand for incident, diffraction, and radiation, respectively; HS stands for hydrostatic and q for quadratic.

In the force equations, the radiation potential has a term involving body-acceleration and another term with body-velocity, which eventually become the added-mass and radiation damping of the system, respectively. The hydrostatic forces, on the other hand, contribute to the stiffness of the structure in the z-direction. For a second-order formulation of the diffraction theory, this leaves the exciting forces on the structure as

$$F_{ex}^{(1)} = F_I^{(1)} + F_D^{(1)} \tag{6-6}$$

$$F_{ex}^{(2)} = F_I^{(2)} + F_D^{(2)} + F_q^{(2)} \tag{6-7}$$

Forces on Fixed Bodies (Diffraction)

Inserting appropriate expressions for potentials into the force-equations, the first-order and second-order difference-frequency forces are obtained in explicit form. The first-order exciting force is

$$F_{ex}^{(1)} = \rho A \sum a_i \{4/(\pi(k_iR)^2 H_1'(k_iR))\} (\omega_i^2/k_i)(1-f_i) e^{i\theta i} \tag{6-8}$$

where $H_1'(k_iR)$ is the derivative of the first-order Hankel function.

Equation 6-8 is often used to compare diffraction theory with Morison's equation and to check the validity of the latter. Clearly, the equation for first-order excitation force is very similar to the first-order inertia force in Morison's equation, which is $= \rho AC_I \sum a_i(\omega_i^2/k_i)(1-f_i)e^{i\theta i}$, with only the factor $4/\{\pi(k_iR)^2 |H_1'(k_iR)|\}$ used here instead of C_I. This factor converges to 2.0 for small values of k_iR (the diffraction parameter), as $|H_1'(k_iR)| \rightarrow 2/\{\pi(k_iR)^2\}$ in these cases. However, such comparisons are not valid for nonlinear forces, as will be shown later.

Ignoring first the structural displacements (deformed geometry and radiation), the other diffraction forces (second-order difference-frequency) are

$$F_I^{(2)} = \pi\rho R \sum\sum \omega_- \chi_{ij} J_1(k_-R) \tanh k_-d\,(1-f_-)/k_-\, e^{i\theta-} \tag{6-9}$$

$$F_D^{(2)} \cong F_{BI}^{(2)} = -\pi\rho R \sum\sum \omega_- \chi_{ij} J_1'(k_-R) \{\tanh k_-d\,(1-f_-) H_1(k_2r)/H_1'(k_2R)/k_2$$

$$+ \sum \tan k_{2e}d\,(1-f_{2e}) K_1(k_{2e}r)/K_1'(k_{2e}R)/k_{2e}\} e^{i\theta-} \tag{6-10}$$

$$\chi_{ij} = -iga_*[k_i^2 \operatorname{sech}^2 k_id/\omega_i - k_j^2 \operatorname{sech}^2 k_jd/\omega_j + 2k_*\{1+(\tanh kd)_*\}\omega_-/\omega_*]/(\omega_-^2/g - k_- \tanh k_-d)$$

$F_{BI}^{(2)}$ is the inertial body force that has been used here as an approximation for $F_D^{(2)}$ for computational efficiency. Previous studies (for example, Eatock Taylor and Jeffreys 1985, Kim and Yue 1990, Kim and Chen 1994) show this to be a reasonable assumption.

The quadratic force for fixed bodies is

$$F_{q0}^{(2)} = F_{FS}^{(2)} + F_{CA}^{(2)} \tag{6-11}$$

Here $F_{FS}^{(2)}$ and $F_{CA}^{(2)}$ denote contributions from the free surface fluctuation (involving $\eta_r^{(1)}$) term and the velocity-squared (involving $\nabla\phi^{(1)}$) term, respectively. They are equivalent to the free surface and convective-axial divergence terms in ME and are given by

$$F_{FS}^{(2)} = -\, i\, 4\rho g/A \sum\sum a_i\, a_j /(k_i\, k_j)\, \Omega_{n-} e^{i\theta-} \tag{6-12}$$

$$F_{CA}^{(2)} = i\, 4\rho g/A\sum\sum a_i a_j\{gd/(\omega_i\omega_j)\}\Omega_{n-}\{I^- + I^+ n(n+1)/k_i k_j R^2\}\cosh k_j d/\cosh k_i d\; e^{i\theta-} \tag{6-13}$$

$$\Omega_{n-} = 1/H'_{n+1}(k_iR)\, H_n'^*(k_jR) - 1/H'_n(k_iR)\, H_{n+1}'^*(k_jR)$$

$$I^{\pm} = \{(\sinh k_+d - \sinh k_+(d-L))/\, k_+d \pm (\sinh k_-d - \sinh k_-(d-L))/k_-d\}/2$$

The centroidal moments are similarly obtained, for example, to get the first-order moment about the spar's CG, only $(1-f_i)$ in Eq. 6-8 should be replaced by $\{-z_{cg} + (z_{cg}+L)\, f_i - (\cosh k_i d - \cosh k_i(d-L))/\sinh k_i d/k_i\}$.

Forces Due to Structural Displacements (Radiation and Deformed Geometry)

In Eq. 6-5, the component $F_q^{(2)}$ is the summation of various quadratic terms in the second-order diffraction theory. Among them are some terms that include structural displacements. For a second-order formulation, the displacements in these terms are of first order. In these quadratic terms, they are multiplied by other first-order terms (forces or displacements) and result in second-order forces. Thus the equation for $F_q^{(2)}$ becomes

$$F_q^{(2)} = -\rho\iint[\,|\nabla\phi^{(1)}|^2/2 + (\xi^{(1)} + \alpha^{(1)}\times x).\nabla\partial\phi^{(1)}/\partial t]n\, dS - \rho g\int \eta_r^{(1)\,2}/2\; n\, dl + \alpha^{(1)}\times F^{(1)} \tag{6-14}$$

In the two dimensional case, the forces due to geometric deformation of the structure are

$$F_G^{(2)} = -\,\rho\iint(\xi^{(1)} + \alpha^{(1)}\times x).\nabla\partial\phi^{(1)}/\partial t\; n dS + \alpha^{(1)}\times F^{(1)} \tag{6-15}$$

In addition to the force from the deformed geometry, the structural displacements contribute to the excitation forces due to radiation. These contributions appear indirectly in the quadratic terms in $F_q^{(2)}$ involving $\eta_r^{(1)}$ and $|\nabla\phi^{(1)}|$. They can be obtained in explicit algebraic form using Eq. 6-14.

Comparison of Diffraction Theory and Morison's Equation

Slender-Body Approximation of Diffraction Forces

So far, the diffraction theory and Morison's equation have been formulated independently using two completely different approaches. In the literature, they have both been used frequently without any theoretical comparison except in the case of first-order potential/inertia forces. As mentioned before, Kim and Chen have in fact shown such correspondence numerically (for an ALP), but there has not been an analytical comparison between the two.

To do that, the nonlinear forces obtained from diffraction theory are derived in case of a slender-body, that is, in the limiting case when the diffraction parameter (kR) tends to 0. In this region, the Bessel functions take the following limiting values

$$J_n(x) \rightarrow (x/2)^n/(n)! \qquad J'_n(x) \rightarrow (x/2)^{n-1}/2(n-1)!$$

$$Y_0(x) \rightarrow 2(\ln(x/2) + 0.577215)/\pi \qquad Y'_0(x) \rightarrow 2/\pi x$$

$$Y_n(x) \rightarrow -(n-1)!(2/x)^n/\pi \qquad Y'_n(x) \rightarrow (n)!(2/x)^n/\pi x \qquad [n \geq 1]$$

$$H_n(x) \rightarrow iY_n(x) \qquad H'_n(x) \rightarrow iY'_n(x),$$

Using these expressions (with γ_{ij} and Γ^- as before), the diffraction forces are simplified to

$$F_{ex}^{(1)} = -\,i\, 2\rho A\, \Sigma(\omega_i^2/k_i)\,(1-f_i)\, e^{i\theta i} \tag{6-16}$$

$$F_I^{-(2)} = -\,\rho A\, \Sigma\Sigma\omega_- \chi_{ij} \tanh k_- d\, (1-f_-)/2\, e^{i\theta-} \tag{6-17}$$

$$F_{BI}^{-(2)} = -\,\rho A\, \Sigma\Sigma\omega_- \chi_{ij} \tanh k_- d\, (1-f_-)/2\, e^{i\theta-} \tag{6-18}$$

$$F_{FS}^{-(2)} = -\,i\,\rho g A\, \Sigma\Sigma a_i\, a_j\, (k_i-k_j)\, e^{i\theta-} \tag{6-19}$$

$$F_{CA}^{-(2)} = i\,\rho g A\, \Sigma\Sigma a_i\, a_j\, (k_i-k_j)\{gd\, k_i k_j/\omega_i\omega_j\, (\Gamma^- + \Gamma^+/2)/\cosh k_i d \cosh k_j d\}\, e^{i\theta-} \tag{6-20}$$

$$F_{G1}^{-(2)} = i\,\rho A/2\, \Sigma\Sigma a_i\, \omega_i^2$$

$$\{(1-f_i)x_{Gj}+(-z_{cg}+(z_{cg}+L)f_i - (\cosh k_i d - \cosh k_i(d-L))/\sinh k_i d/k_i)\alpha_j\}\, e^{i\theta-} \tag{6-21}$$

$$F_{G2}^{(2)} = \alpha_2^{(1)}\, F_3^{(1)}\, i - \alpha_2^{(1)}\, F_1^{(1)}\, k \tag{6-22}$$

The radiation forces tend to zero in this limiting case of slender-body approximation. Therefore, they do not contribute to the second-order forces here.

Using $K_I = \rho C_I A \to 2\rho A$ and $K_m \to \rho A$ (assuming $C_I = 2$), the horizontal forces and moments about CG (for deep water) are shown in Tables 6-3 and 6-4. The comparison with Tables 6-1 and 6-2 is interesting. The first-order forces, as well as the second-order potential, free-surface, and convective forces are equal in both cases. However, the corresponding forces from axial divergence (without deformations) and structural displacement are twice and one-fourth as much as before, respectively.

Table 6-3. Difference-Frequency Diffraction Forces ($d \to \infty$, $kR \to 0$)

Type of Force	Resultant Force
First-order potential	$gK_I \Sigma a_i [r_i] \sin\theta_i$
Second-order potential (I,BI)	$gK_I \Sigma\Sigma a_* k_i (1-\omega_j/\omega_i)[r_-] \sin\theta_-$
Convective	$-K_I \Sigma\Sigma a_* \omega_*(k_-/k_+)[r_+] \sin\theta_-$
Axial divergence	$K_m\Sigma\Sigma a_* \omega_* \{(r_+) k_-/k_+\}(\sin\theta_-)$
Free-surface fluctuation	$gK_I \Sigma\Sigma a_* k_-(\sin\theta_-)/2$
Structural surge	$gK_I \Sigma\Sigma \{x_{Gim}^{(1)} a_j^{(1)} k_j[r_j] - x_{Gjm}^{(1)} a_i^{(1)} k_i[r_i]\}(\sin\theta_-)/8$
Structural rotation	$gK_I \Sigma\Sigma \{\gamma_{im}^{(1)} a_j^{(1)} k_j [d_j] - \gamma_{jm}^{(1)} a_i^{(1)} k_i [d_i]\}(\sin\theta_-)/8$

Note: $r_i = 1-e^{-k_iL}$, $R_i = 1+e^{-k_iL}$, $d_i = L R_i/2 - r_i/k_i$, and such assuming $z_{cg} = -L/2$

Table 6-4. Difference-Frequency Diffraction Moments ($d \to \infty$, $kR \to 0$)

Type of Moment	Moment about CG
First-order potential	$gK_I \Sigma a_i \{d_i\} \sin\theta_i$
Second-order potential (I,BI)	$gK_I \Sigma\Sigma a_* k_i (1-\omega_j/\omega_i)\{d_-\}\sin\theta_-$
Convective	$-K_I \Sigma\Sigma a_* \omega_*(k_-/k_+)\{d_+\} \sin\theta_-$
Axial divergence	$K_m\Sigma\Sigma a_* \omega_* \{(d_+)k_-/k_+\}(\sin\theta_-)$
Free-surface fluctuation	$gLK_I \Sigma\Sigma a_* k_-(\sin\theta_-)/4$
Structural surge	$gK_I \Sigma\Sigma \{x_{Gim}^{(1)} a_j^{(1)} k_j d_j - x_{Gjm}^{(1)} a_i^{(1)} k_i d_i\}(\sin\theta_-)/8$
Structural rotation	$gK_I\Sigma\Sigma \{\gamma_{im}^{(1)} a_j^{(1)}[(k_jL^2/4+2/k_j)r_j - LR_j] -$ $\gamma_{jm}^{(1)} a_i^{(1)}[(k_iL^2/4+2/k_i)r_i - LR_i]\}(\sin\theta_-)/8$

Note: $r_i = 1-e^{-k_iL}$, $R_i = 1+e^{-k_iL}$, $d_i = L(1+e^{-k_iL})/2-(1-e^{-k_iL})/k_i$, etc.; assuming $z_{cg} = -L/2$

Morison's equation is commonly believed to be the slender-body equivalent of the diffraction theory and is frequently used for slender bodies with small diffraction parameters based on first-order inertia forces. The results of this section show the limitations of this assumption for second-order difference-frequency forces and moments.

Therefore, Morison's equation (or the extension by Rainey) is not referred to as the slender-body theory in subsequent discussions. Instead, a more specific name, Slender Φ2 theory, is used for the formulation based on the forces and moments derived in this section. The validity of this formulation and its limitations in representing the second-order effects is discussed while comparing the numerical results.

Newman's Approximation

The previous section shows some differences between the forces calculated from diffraction theory and Morison's equation in the limiting case when the diffraction parameter tends to zero. However, this limiting condition is not applicable for real structures. At very small frequencies, the difference-frequency forces from diffraction theory tend to diverge from the slender-body approximation.

In Kim and Chen's work, the very low-frequency forces using slender-body theory and diffraction theory were quite different. Apart from the mean-drift force, this may result in significant difference in resonant response of very flexible structures.

Newman's (1974) approximation (also Marthinsen 1983) is the other simplified limiting case of diffraction theory. It suggests that the quadratic force transfer function (QTF) can be approximated by its corresponding diagonal value. The total force can be written in terms of QTF (Q_{ij}) as,

$$F_{-}(t) = \Sigma\Sigma a_{*} (\{Q_{ij}\}_R \cos\theta_{-} + \{Q_{ij}\}_I \sin\theta_{-}) \qquad (6\text{-}23)$$

where the summations are for whole ranges of i and j.

Newman's approximation suggests, $\{Q_{ij}\}_R = \{Q_{ji}\}_R \cong \{Q_{ii} + Q_{jj}\}_R /2$; $\{Q_{ij}\}_I = \{Q_{ji}\}_I \cong 0$. In time domain, this is further simplified to

$$F_{-}(t) \cong 2\,(\Sigma a_i \sqrt{\{Q_{ii}\}_R} \cos\theta_i)^2 \qquad (6\text{-}24)$$

For irregular waves with many wave components, this approximation simplifies the computational task considerably. Instead of calculating Q_{ij} at each combination of frequencies (i,j), it only needs to be computed at frequencies (i,i). The forces predicted by Newman's approximation are easily obtained from the second-order forces using diffraction theory. Only the quadratic forces $F_q^{-(2)}$, radiation forces $F_r^{-(2)}$, and forces due to structural displacements contribute here. The other forces, such as $F_I^{-(2)}$ and $F_{BI}^{-(2)}$, vanish at zero difference-frequency. The transfer functions due to forces $F_q^{-(2)}$ and $F_r^{-(2)}$ are shown in Eqs. 6-25 through 6-28.

The quadratic transfer functions from free-surface fluctuation (FS) and convective + axial divergence (CA) are as follows:

$$Q_{FSii}^{(2)} = 4\rho g R / \pi (k_i R)^2 \, \mathrm{Im}(1/H'_{n+1}(k_i R)\, H_n'^{*}(k_i R)) \qquad (6\text{-}25)$$

$$Q_{CAii}^{(2)} = -\omega_i^2 / \{g\,(\sinh k_i d)^2\} Q_{FSii}^{(2)} \{J^{-} + J^{+} n(n+1)/(k_i R)^2\} \qquad (6\text{-}26)$$

where $J^{\pm} = \{(\sinh 2k_i d - \sinh 2k_i(d-L))/2k_i \pm L\}/2$ have been used.

The Newman forces due to displaced geometry are not shown. These are the same as shown for the Φ2 theory, with j = i and the force having an extra factor of one-half. The contributions from radiation forces are also due to free-surface fluctuation and convective + axial divergence. They are

$$Q_{rFii}^{(2)} = -(\pi\rho/2R)\,\mathrm{Re}\{C_i^2\,F_{2i}\,A_i^* + \Sigma F_{2i}\,C_{ei}^*\,Cc_{ii+}\}/a_i^2 \tag{6-27}$$

$$Q_{rCii}^{(2)} = -(\rho R/2)\,\mathrm{Re}\{S_i^2\,G_i\,B_i^* + \Sigma G_i\,D_{ei}^*Cc_{ii-}\}/a_i^2 \tag{6-28}$$

where $C_i^2 = J^+$, $S_i^2 = J^-$, $Cc_{ii\pm} = \{(\sinh K_+d - \sinh K_+(d-L))/K_+ \pm (\sinh K_-d - \sinh K_-(d-L))/K_-\}/2$ and such. $[K_\pm = k_i \pm i\kappa_{ei}]$ and the other symbols are as defined before.

Similar expressions can be derived for moments about the CG.

One significant aspect here is that all the force components are real; that is, they are associated with cosines only. This makes them the principal component of mean-drift and low frequency forces. Obviously, Newman's approximation is more accurate at small frequency-differences (predicts the mean-value exactly), and its accuracy deteriorates at higher frequency-differences, where the imaginary components (better predicted by Morison's equation) become more prominent.

Combining Newman's Approximation With Slender-Body Φ2 Theory

In Morison's equation, all the inertia-forces are associated with sine terms; that is, the forces are all imaginary terms. On the other hand, the diffraction-forces have complex values, with real as well as imaginary components. These real-value components are not taken into account in the slender-body approximation. Yet, this can be important particularly at small frequencies. For example, the inertia force in Morison's equation has no mean-value component, and the mean-drift forces in slender-body formulation are taken from the diffraction theory.

For arbitrarily slender structures, the slender-body theory should still be a good approximation of Φ2 theory, while for arbitrarily small frequencies, Newman's approximation is more accurate. But for practical structures with non-zero dimensions and non-zero natural frequencies, these two approximate methods are not always reliable.

The inherent lack of consistency between Morison's equation and Φ2 theory is illustrated in Anam & Roësset (2004a, 2004b). This work shows that the individual nonlinearities may result in different response phases even if their amplitudes are similar. Therefore, when acting together, their resultants can be quite different. For example, the effects of convective acceleration and free surface fluctuation are found to be equal and opposite using Morison's equation. However, no such conclusion can be drawn from the Φ2 theory.

A similar comparison can also be made between Newman's approximation and the Φ2 theory. The former is based solely on the amplitude of the quadratic transfer function, without any consideration to the phase. Therefore, the individual nonlinear effects do not interact in a manner similar to Φ2 theory, and the resultant discrepancy can be quite significant at higher frequencies. The nonlinear effects in Newman's approximation and their interactions with each other are quite different from either the Φ2 theory or Morison's equation. An interesting result is that the free surface fluctuation and the convective-axial divergence terms are almost equal and additive for

Newman's approximation, which is the opposite conclusion from Morison's equation. In Φ2 theory, they interact quite differently.

To bridge the gap between the basic theoretical discrepancies, a new combined-force approach is suggested in this work. The objective is to model the individual second-order effects from Φ2 theory without the rigorous calculations involved. It should also provide some insight into the Φ2 theory results. As mentioned before, the inertia portion of Morison's equation (both full-time domain and second order) predicts very small forces at small frequencies and a zero mean-drift force (except drag) at zero frequency. Thus, it diverges from the Φ2 theory results at small frequencies. Newman's approximation, on the other hand, is very accurate at small frequencies but loses its accuracy at larger frequencies. In the combined-force approach, these two simple methods are combined in order to simulate Φ2 theory results. Instead of Morison's equation, the slender-body approximation of Φ2 theory is used as a force component of this new model because it is a limiting case of the Φ2 theory for extremely slender structures. The forces from this approach were derived explicitly in the previous section (also shown in Tables 6-3 and 6-4), and its implementation is as convenient as the second-order Morison's equation.

The rationale for the combined-force method is the fact that a significant shortcoming in the slender-body approximation of Φ2 theory is the total omission of mean-drift forces and, in general, the underestimation of slow-drift forces. In this range of low frequencies, the Φ2 theory converges to Newman's approximation. The combined-force method is the combination of two limiting cases.

Numerical Results

Bichromatic Wave

To illustrate the analytical results numerically, a model spar (with diameter about 40 m, draft nearly 200 m) is first subjected to a bichromatic wave consisting of two waves each of 3.0 m amplitude acting at frequencies ω_1 = 0.442 rad/sec and ω_2 = 0.524 rad/sec, so that the second-order difference-frequency forces act at a frequency $\omega_2 - \omega_1$ = 0.082 rad/sec. The spar displacements are calculated here by ME, ME2, Φ2 theory, slender body Φ2 theory, Newman's approximation, and the combined-force method. The main objective is to investigate the difference among the results, both in terms of magnitudes and constituents of the individual nonlinear effects.

The results for the second-order surge responses (magnitude and phase) are presented in Table 6-5, and the results for the second-order pitch in Table 6-6. The results from the slender Φ2 theory approach are qualitatively similar to the second-order Morison's equation results, although the effects of displaced geometry are nearly three to four times smaller here (as indicated by the corresponding expressions from Tables 6-1 and 6-2). The second-order effects in the slender body theory represent almost co-linear vectors in the complex plane. However, this is not the case for the displacements from Φ2 theory, where the hydrodynamic forces are obtained in the form of complex Bessel/Hankel functions. The combined-force method represents these results quite well. It is clear that although some differences exist, the individual nonlinear effects from the Φ2 theory (both the magnitudes and phases) are now better represented.

Table 6-5. Nonlinear Surge Response [Amplitude (m), Phase (deg)]

	ME	ME2	Φ2 Theory	Slender Φ2	Newman	Combined
Temp2	0.270, 48.15	0.270, 48.15	0.270, 48.15	0.270, 48.15	*	0.270, 48.15
ConAx	0.317, 229.37	0.315, 229.12	0.336, 207.63	0.247, 229.27	0.119, 140.13	0.275, 203.77
FS	0.329, 50.71	0.344, 50.90	0.311, 68.13	0.344, 50.90	0.135, 139.49	0.373, 72.20
DisG	0.308, 51.79	0.331, 48.55	0.110, 21.90	0.087, 51.55	0.052, 318.09	0.099, 19.54
Total	0.590, 50.82	0.630, 49.38	0.388, 75.07	0.454, 50.28	0.202, 140.23	0.497, 74.23

Using the combined-force method, the surge displacements due to quadratic terms (convective and axial), free surface fluctuation, and displaced geometry are [0.275 m, 203.77°], [0.373 m, 72.20°], and [0.099 m, 19.54°], respectively. They compare better with the corresponding Φ2 theory displacements, that is, [0.336 m, 207.63°], [0.311 m, 68.13°], and [0.110 m, 21.90°], respectively. The slender Φ2 theory predicts them to be [0.247 m, 229.27°], [0.344 m, 50.90°], and [0.087 m, 51.55°], respectively, which represent almost co-linear terms (with phase angles of around 50° or 230°, which are either in-phase or 180° out of phase) in the complex plane unlike the displacements predicted by the Φ2 theory. The results from ME and ME2 are similar to the slender Φ2 theory, except the effects of displaced geometry. The second-order temporal acceleration (Temp2) is represented almost identically [0.270 m, 48.15°] by all these methods. Newman's approximation, on the other hand, has only three non-zero nonlinear effects, that is, the effect of second-order potential is now zero. Similar comparisons can be made for the pitch displacements.

Table 6-6. Nonlinear Pitch Response [Amplitude (deg), Phase (deg)]

	ME	ME2	Φ2 Theory	Slender Φ2	Newman	Combined
Temp2	0.207, 230.09	0.207, 230.09	0.207, 229.91	0.207, 230.09	*	0.207, 230.09
ConAx	0.674, 50.99	0.589, 50.93	0.751, 24.38	0.519, 50.94	0.367, 320.85	0.635, 15.61
FS	0.844, 232.16	0.881, 232.47	0.798, 249.29	0.881, 232.47	0.345, 320.87	0.956, 253.65
DisG	0.670, 233.51	0.844, 230.14	0.225, 208.39	0.189, 233.62	0.088, 139.61	0.203, 207.96
Total	1.047, 233.37	1.344, 231.31	0.705, 275.78	0.759, 233.15	0.625, 321.04	1.000, 271.78

The resultant second-order response amplitudes and phases for the combined-force method are [0.497 m, 74.23°] for surge and [1.000°, 271.78°] for pitch, compared to the Φ2 theory amplitudes and phases of [0.388 m, 75.07°] and [0.704°, 275.78°]. The responses are [0.590 m, 50.82°] and [1.047°, 233.37°] using ME and [0.630 m, 49.38°] and [1.344°, 231.31°] using ME2, while the slender Φ2 theory responses are [0.454 m, 50.28°] and [0.759°, 233.15°], respectively, and those from Newman's approximation are [0.202 m, 140.23°] and [0.625°, 321.04°].

Although both response amplitudes from the combined-force method are higher (by about 30 to 40 percent) than the corresponding Φ2 theory results, the phases are very well predicted. On the other hand, the slender Φ2 theory amplitudes are closer to the Φ2 theory amplitudes in this case, but they fail to represent properly the real and imaginary components (the phases) of the displacements. Another noticeable result is the difference between the ME and ME2 responses, particularly for pitch. This shows the presence of third- and higher-order forces in ME, which cannot be represented by a second-order formulation like ME2.

These discrepancies would be more evident in analyzing irregular waves because there the resultant amplitudes are obtained from several individual interactions so that both the amplitude and phase of the individual effects are important.

Irregular Wave JONS

The results from different force models are compared here for a random wave with the JONSWAP spectrum (called JONS here) with significant wave height of 13 m and a peak period of 14 seconds. For this case, the surge response at the mean water level (MWL) from the ME is presented in Figure 6-1. The main interest is in the resonant surge and pitch response. The first peak in the figure indicates the resonant surge response (at the CG of the spar), while the second peak is due to the resonant pitch (times the distance from CG to the MWL). The numerical values of these peak amplitudes are 3.074 m and 3.217 m, respectively.

The resonant responses from the Φ2 theory (Fig. 6-3) are (3.905 m, 1.852 m), which are used as reference for comparison. The results from ME2, the Φ2 theory, slender Φ2 theory, Newman's approximation, and the combined-force approach are presented in Figures 6-2 to 6-6. The peak amplitudes from ME2 are [3.348 m, 3.966 m]. Both the ME and ME2 underestimate the resonant surge response (by 15 to 20 percent) and highly overestimate the pitch amplitude (by about 75 to 120 percent). The differences between the ME and ME2 results are also noticeable, particularly for pitch.

The responses from the slender Φ2 theory are (2.260 m, 2.172 m). Thus, the resonant amplitude at the pitch natural frequency is reasonably well predicted (less than 20 percent over-estimated), but the resonant surge amplitude is significantly (more than 40 percent) underestimated. The peak responses from Newman's approximation are [3.790 m, 1.329 m], which are both less than the Φ2 theory results, by about 3 percent and 30 percent, respectively. This shows the gradual divergence of this method from the Φ2 theory at higher frequencies.

The combined-force approach, on the other hand, predicts resonant responses of [4.141 m, 2.371 m], respectively. Both the surge and pitch amplitudes are overestimated by this method, as was the case for the bichromatic wave. Although the overestimation for surge is only about 6 percent, the pitch is nearly 30 percent overestimated.

A closer look at the responses near the surge and pitch resonant regions (Fig. 6-7) show that in addition to the peak responses, the combined-force method also captures the general trend of the Φ2 theory results better than Newman's approximation and the slender Φ2 theory. The response spectra from the Φ2 theory and the combined-force method are very similar, which cannot be said of the other methods. For example, the ordinates closest to the peak amplitudes are (2.226 m, 1.663 m) from the Φ2 theory, (2.657 m, 1.921 m) from ME, (2.860 m, 3.058 m) from ME2, (1.972 m, 1.595 m) from the slender Φ2 theory, (3.135 m, 0.411 m) from Newman's approximation, and (2.155 m, 1.648 m) from the combined-force method. These results and Figure 6-7 in general show further how closely the combined-force method represents the Φ2 theory results. In this case, Newman's approximation models the Φ2 theory results well near the surge natural frequency, where slender Φ2 theory fails to capture the general trend. On the other hand, the latter performs better near the pitch natural frequency, while Newman's approximation fails. Neither ME nor ME2 captures the general trend of response appropriately near the natural frequencies. However, only the combined-force model is satisfactory around both frequencies.

The computational efficiency provided by the combined-force model compared to the Φ2 theory can be considerable when dealing with irregular waves with a large number of wave components. For example, the JONSWAP spectrum in the present case has 700 wave components. The calculation of difference-frequency forces by the Φ2 theory required 115 seconds with a 1.1 GHz Pentium3 Celeron processor. Similar calculation required less than 1 second when using the combined-force model. The computational effort for the combined-force model is slightly more than Newman's approximation and Morison's equation.

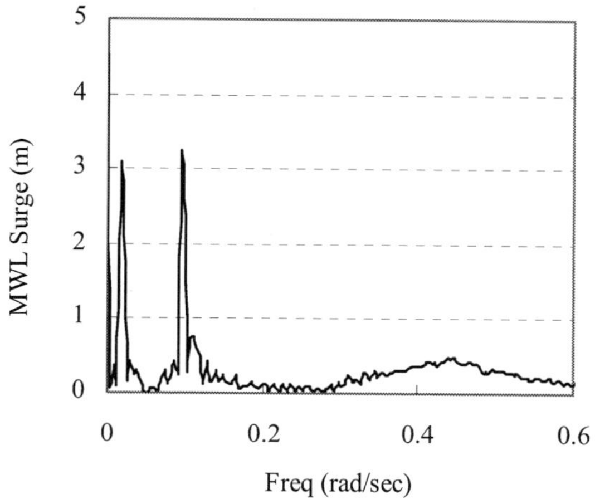

Fig. 6-1. JONS: MLW surge (ME)

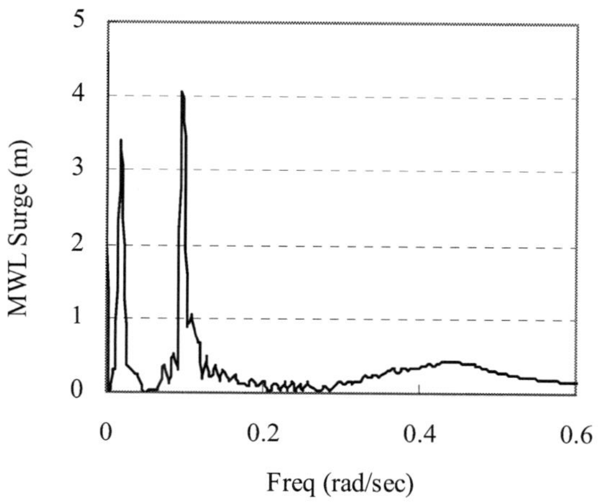

Fig. 6-2. JONS: MLW surge (ME2)

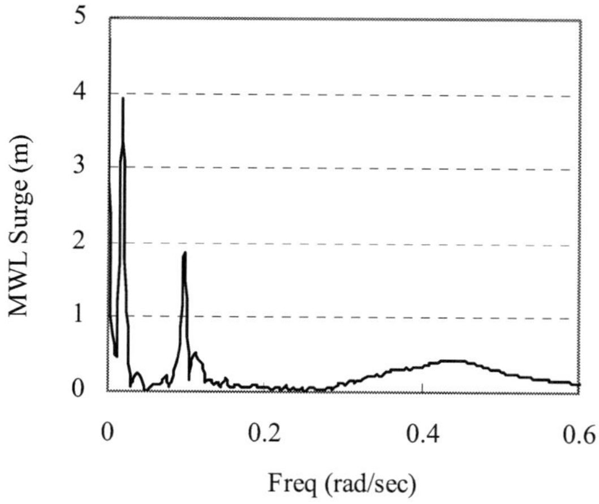

Fig. 6-3. JONS: MLW surge (Diff2)

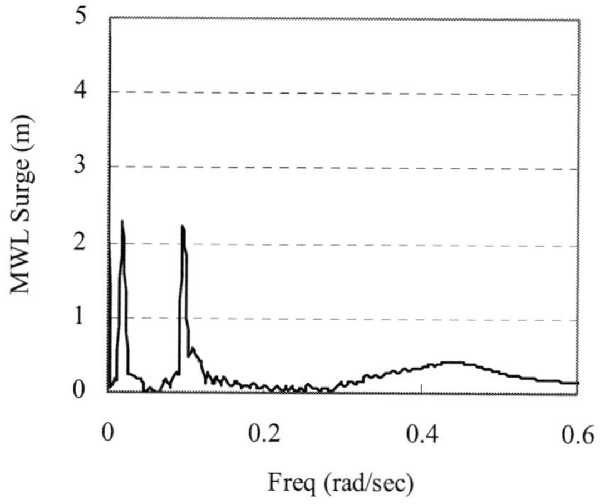

Fig. 6-4. JONS: MLW surge (Slender Diff2)

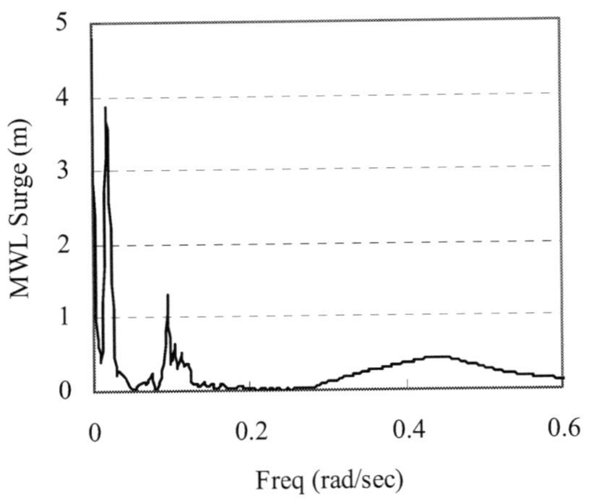

Fig. 6-5. JONS: MLW surge (Newman)

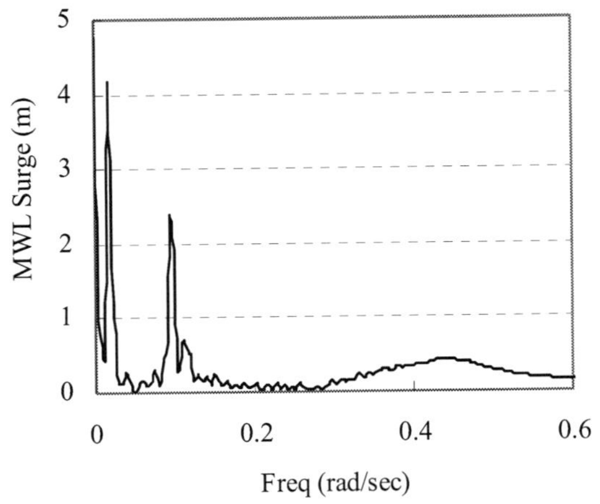

Fig. 6-6. JONS: MLW surge (combined)

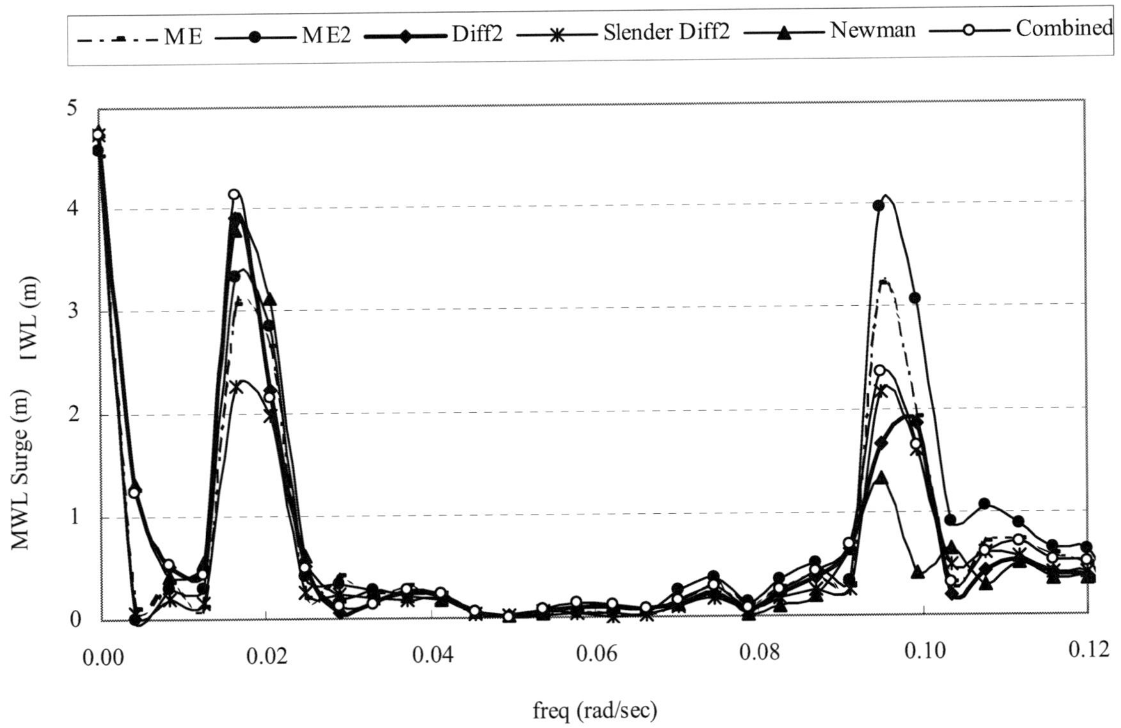

Freq (rad/sec)

Fig. 6-7. Enlarged low-frequency portions of Figs. 6-1 to 6-6

As previously mentioned, the most important aspect of the combined-force approach is that it provides a better understanding of the results expected from the Φ2 theory and suggests a conceptual bridge between existing simple force models like Newman's approximation and slender body theory.

Conclusions

This chapter compares in detail various types of nonlinear responses for spar platforms predicted by different hydrodynamic force models. The second-order difference-frequency forces and responses are primarily compared, illustrating their magnitudes and phases in the complex plane. The conclusions are summarized as follows:

The second-order effects from all the slender body formulations have corresponding nonlinear effects in the Φ2 theory. Although many of the nonlinear forces have similar magnitudes in both formulations (except the forces due to displaced geometry and axial divergence), their constituents in the complex plane can be quite different. Their discrepancy at small frequencies can be very significant for flexible platforms like spars.

The second-order Morison's equation (ME2) represents the full-time domain Morison's equation (ME) reasonably well, but the third- and higher-order effects may not be negligible in some cases.

The second-order difference-frequency forces from the slender Φ2 theory follow the general pattern (the phase angles) of the corresponding forces from Morison's equation, but their numerical values do not always match. This is particularly true for the forces due to displaced geometry. Therefore, contrary to common belief, Morison's equation is not equivalent to the slender-body approximation of the diffraction theory for this class of nonlinear problems.

The combined-force method provides a better insight into the nonlinear difference-frequency effects involved in the second-order diffraction theory and represents the constituents of the nonlinear effects better than the existing approximations of the diffraction theory.

References

Anam, I. (2000). "Evaluation of the dynamic response of spar platforms," Ph.D. Dissertation, Texas A&M University, College Station, TX, USA.

Anam, I., and Roësset, J. M. (2004). "Nonlinear Hydrodynamic Forces on Flexible Structures: A Combined-Force Model." *ASME J. of Offshore Mech. & Arctic Engrg.*, 126(1), 78-83.

Anam, I., and Roësset, J. M. (2004). "Slender-Body Approximations of Hydrodynamic Forces for spar Platforms." *Intl. J. of Offshore & Polar Engrg.*, 14(2), 104-109.

Cao, P., and Zhang, J. (1996). "Slow motion responses of compliant offshore structures." *Proc. 6th Intl. Symp. Offshore & Polar Engrg.*, Los Angeles, CA. USA, 1, 296-303.

Eatock Taylor, R., and Jeffreys, E. R. (1985). "Validity of hydrodynamic load predictions for a Tension Leg Platform," *Ocean Engrg.*, 13, 449-490.

Jha, A. K., de Jong, P. R., and Winterstein, S. R. (1997). "Motions of a spar buoy in random seas: Comparing predictions and model test results." *Beh. Offshore Str.-97*, 2, 333-347.

Kim, M. H., and Chen, W. (1994). "Slender-body approximation for slowly-varying wave loads in multi-directional waves," *Appl. Ocean Res.*, Vol. 16, 141-163.

Kim, M. H., and Yue, D. K. P. (1990). "The complete second-order diffraction solution for an axisymmetric body: Part 2. Bichromatic incident waves and body motions," *J. Fluid Mech.*, 211, 557-593.

MacCamy, R. C., and Fuchs, R. A. (1954). "Wave forces on piles: A Diffraction Theory," Tech. Memo. 69, Beach Erosion Board.

Marthinsen, T. (1983). "Calculation of slowly varying drift forces." *Appl. Ocean Res.*, 5, 141-144.

Mekha, B. B., Johnson, C. P., and Roësset, J M. (1995). "Nonlinear response of a spar in deep water: Different hydrodynamic and structural models." *Proc. 5th Intl. Symp. Offshore & Polar Engrg.*, The Hague, The Netherlands, III, 462-469.

Mekha, B. B., Weggel, D. C., Johnson, C. P., and Roësset, J M. (1996). "Effects of second order diffraction forces on the global response of spars." *Proc. 6th Intl. Symp. Offshore & Polar Engrg.*, Los Angeles, CA, USA, I, 273-280.

Morison, J. R., O'Brien, M. P., Johnson, J. W., and Shaaf, S. A. (1950). "The force exerted by surface waves on piles," *Pet. Trans*, 189, 149-157.

Newman, J. N. (1974). "Second order slowly varying forces on vessels in irregular waves." *Symp. Dynamics of Marine Vehicles & Structures in Waves*, University College, London, UK.

Rainey, R. C. T. (1989). "A new equation for calculating wave loads on offshore structures," *J. Fluid Mech.*, 204, 295-324.

Ran, Z., Kim, M. H., Niedzwecki, J. M., and Johnson, R. P. (1996). "Responses of a spar platform in random waves and currents (Experiment vs. Theory)." *Intl. J. Offshore & Polar Engrg.*, 6(1), 27-34.

Ran, Z., and Kim, M. H. (1997). "Nonlinear coupled responses of a tethered spar platform in waves." *Intl. J. Offshore & Polar Engrg.*, 7(2), 111-118.

Ran, Z., Kim, M. H., and Zheng, W. (1999). "Coupled dynamic analysis of a moored spar in random waves and currents (Time-domain versus frequency-domain analysis)." *J. Offshore Mech. Arctic Engrg.*, 121, 194-200.

Weggel, D. C., and Roësset, J. M. (1996). "The behavior of spar platforms." *OTRC spar Related Papers*, Offshore Tech. Res. Ctr. (OTRC), Texas A&M University, College Station, TX, USA.

Weggel, D. C., and Roësset, J. M. (1996). "Second order dynamic response of a large diameter spar platform: Numerical predictions versus experimental results." *Offshore Mech. Arctic Engrg. Conf.*, Florence, Italy, I, 489-496.

Chapter 7: Spar Vortex Induced Motion Considerations for Design

By Tim Finnigan, Principle Engineer, Chevron-Texaco

Introduction

Floating offshore structures, which are made up of large cylindrical structures such as spars, semi-submersibles, and TLPs, can respond to vortices shed from those members when exposed to persistent currents. This response is referred to as vortex induced motion (VIM). Because most of the experience in the offshore industry to date has been for spar platforms, this chapter will focus on the VIM behavior of spar platforms.

VIM Fundamentals

When a cylinder is exposed to a steady current, alternating eddies or vortices are formed and shed at a regular period. Figure 7-1 shows how these eddies appear in the downstream wake of a cylinder. This is referred to as vortex shedding.

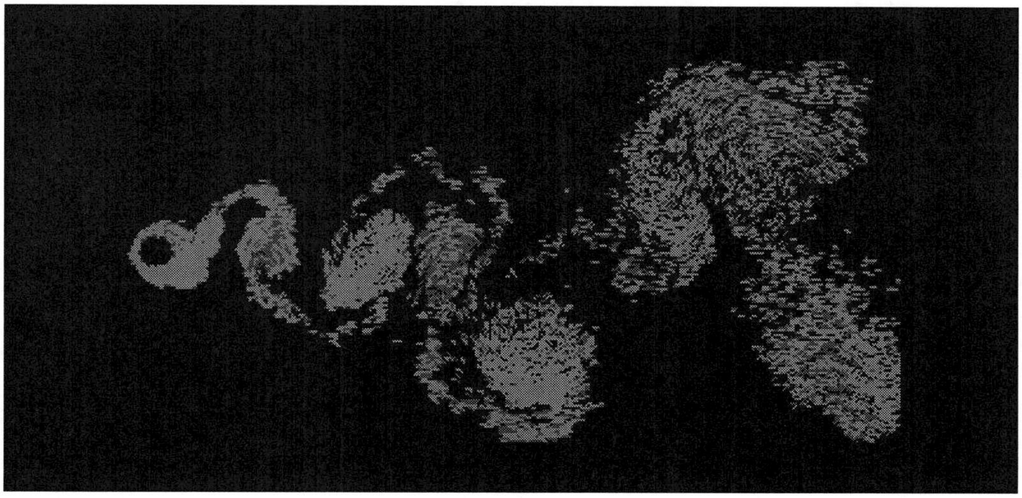

Fig. 7-1. Eddies in the downstream wake of a cylinder

The vortex shedding frequency, f_s , also known as the Strouhal frequency, is related to the non-dimensional Strouhal number, S:

$$f_s = \frac{SV_c}{D}$$

where

V_c = current velocity

D = cylinder diameter

The inverse of this number,

$$T_s = \frac{1}{f_s}$$

referred to as the Strouhal shedding period, is also sometimes used.

When the vortices are shed, they create a pressure variation around the cylinder, which leads to alternating lift and drag forces. When the shedding period is close to a natural period of a structure, oscillations of the structure can occur at the shedding period. This phenomenon is traditionally known as vortex induced vibration (VIV) and is well known for risers and tendons. The tendency for the response to synchronize with the shedding period is known as lock-in because the cylinder tends to lock in to the shedding frequency of the vortices.

VIV can occur for almost any length of cylinder and is not restricted to cylinders of circular cross section. Spar platforms, for example, can experience vortex-induced oscillations when their sway (sideways motion relative to the current) period is close to the Strouhal period. Figure 7-2 shows an example of the motion of a spar in a loop/eddy current of about 2 kts.

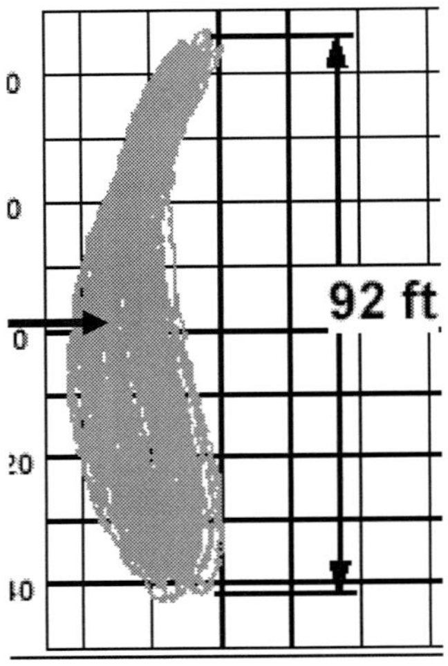

Fig. 7-2. Motion trajectory of a spar experiencing VIM (Source: OTRC/MMS Workshop 2003)

The sway period of the motion for the spar in Figure 7-2 is about 180 seconds, which is close to the natural sway period of the spar. There is also a smaller motion in the in-line direction (surge) at one-half the transverse period. Because the natural period of spars is so long, the term "vortex induced motions" (VIM) is typically used in lieu of VIV, although the phenomenon is the same.

Transverse (Cross Flow) VIM

The occurrence of lock-in is related to the non-dimensional reduced velocity V_{rn},

$$V_{rn} = \frac{V_c T_n}{D}$$

where T_n = natural period in calm water.

Lock-in typically occurs for values of $4 < V_{rn} < 10$ for transverse VIM. Unlike other resonant responses, the amplitude of VIM is bounded.

The transverse motion amplitude (a) is given in non-dimensional units as the ratio of the sway amplitude to diameter (a/D). The largest single amplitude transverse motion observed on bluff bodies is on the order of a/D = 1, although under certain conditions this value may exceed 1.0. The amplitude of response varies for different structures and depends on parameters such as the structural shape, vortex mitigation devices, appurtenances, current profile, mass ratio, and damping. Further, an asymmetrical mooring system could result in a VIM trajectory for which the major axis of the VIM is not transverse to the current direction.

VIM varies with V_{rn} and is negligible when V_{rn} is below a threshold value of approximately 4. The range of V_{rn} where the largest VIM occurs is often referred to as the locked-in range. This typically occurs between the values of 6 and 8 but could be as high as 10, depending on parameters such as the structural shape, vortex mitigation devices, appurtenances, and current profile. Beyond the locked-in range the VIM decreases. This region is sometimes referred to as the lock-out range; however, the value at which VIM locks out is quite variable and can be fairly large for some structures and conditions (for example, $V_{rn} = 14$ or greater). For some geometries, the response may never completely lock out and, in fact, could increase again at much higher reduced velocities due to a phenomenon known as galloping. However, discussion of that phenomenon is beyond the scope of this document.

An example of V_{rn} versus a/D design criteria is provided in Figure 7-3, showing the lock-in slope, locked-in region, and the locked-out region. This type of curve is typically used to define the VIM response amplitude.

For most spars and other moored offshore vessels experiencing VIM, the amplitude of response varies with current direction for the same reduced velocity. However, it is not advisable to take advantage of low-response headings in design, as small geometric changes or changes in heading could lead to a higher response than expected.

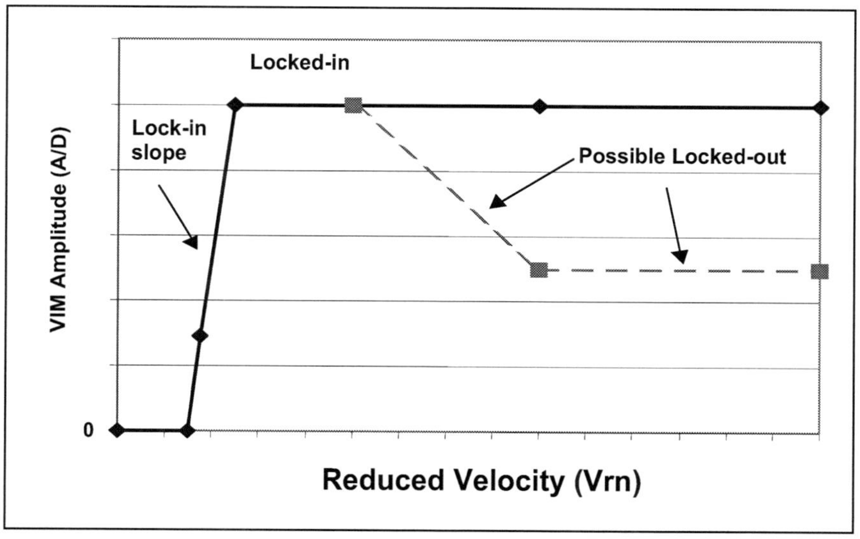

Fig. 7-3. Typical VIM amplitude versus reduced velocity

The structure natural period, T_n, is a function of mooring stiffness and vessel mass:

$$T_n = 2\pi\sqrt{\frac{(M + Ma)}{K}}$$

where
M = mass of spar
Ma = added mass of spar
K = global lateral stiffness at mean offset position

Spar added mass is usually determined by analytical tools or model testing. The mooring stiffness can be significantly different for various offsets, especially for grouped mooring patterns. The transverse stiffness used for calculating V_{rn} is usually evaluated at the mean offset position under the combined current and associated wind and wave loading. However, because the mean offset is dependent on the drag force, which is in turn dependent on the VIM amplitude, the process of selecting the appropriate offset for VIM calculation is iterative. Note that the observed period from model tests or field measurements may be different from the calculated still water natural period. Therefore, when available, field measurements should be used to calibrate the model test data.

Inline VIM

Inline VIM is typically in the direction of the current, and it may affect the transverse VIM amplitude. Inline VIM is also a function of the parameters discussed above for transverse VIM. The magnitude of inline VIM is typically much less than the transverse response. Field measurement data for a classic spar with an equally spaced spread mooring system indicate inline a/D of 10 percent to 15 percent of the transverse a/D (OTRC/MMS Workshop 2003). However, the magnitude can be higher than that if the natural period for the inline motion is close to half of the natural period for the transverse motion (resonance condition).

Drag Coefficient

Model tests are often used to determine the drag coefficient (C_d) to be used in design. A "base drag" C_{d0} is assumed for the case with no VIM (a/D = 0), and an amplification factor, k, is applied to account for VIM effects. This drag augmentation is a function of a/D and V_{rn} and can be expressed as (Edwards et al. 2003, Skip et al. 1977, Vandiver 198):

$$C_d = C_{d0}\left[1 + k\left(\frac{a}{D}\right)\right]$$

Where C_{d0} is the drag coefficient with no VIM and k is the amplification factor that needs to be determined by model tests.

The mean drag force on the cylinder is given by

$$F_d = C_d \frac{1}{2}\rho V_c^2 DL$$

C_d = mean drag coefficient (absolute current velocity) in the presence of VIM

ρ = density of the fluid
L = hull draft
For a sheared current profile, the force would be integrated along the spar hull draft:

$$F_d = C_d \frac{1}{2} \rho \int_{-L}^{0} V_c^2(z) D dz$$

For truss spars, which consist of a large number of components of various diameters, D is defined as the diameter of the hard tank. The drag force on the truss section is computed separately as the sum of drag force on the small diameter truss members. The truss members could induce additional damping when the truss spar undergoes VIM.

Practical Considerations

There are a number of special issues to consider for spar VIM, which is different than response predictions for other environmental loads.

There are presently no validated analytical tools for the prediction of spar VIM. VIM design criteria are usually obtained from model testing. However, computational fluid dynamics (CFD) methods have improved significantly during the past few years and show promise in their ability to predict VIM amplitudes (Halkyard et al. 2005, Oakley 2007).
1. Model testing methods should be validated with field measurement data, but such data are quite limited and difficult and expensive to obtain.
2. Model tests cannot model all environmental effects and all scale laws. Hence, care needs to be exercised in the interpretation and use of model test data.
3. Spar VIM is affected by current velocity, direction, profile, hull geometry, truss details (if any), appurtenances, surface roughness, and turbulence. All of these factors should be considered in the model test program and when applying model test results to design.
4. The duration for peak current and resulting VIM can be much longer than peak storm duration. Appropriate statistical tools are needed to predict extreme VIM amplitudes.
5. The calibration of the factors of safety for mooring design does not include the spar VIM condition and the uncertainties associated with spar VIM. Consequently, sensitivity checks may be warranted.

Because of the above issues, it is important to address VIM conservatively in the spar mooring and riser design stage.

Environmental Considerations

Current velocity, profile, direction, and duration affect vessel VIM. The most common categories of currents are tidal, circulation currents (for example, the Gulf Stream, the Gulf of Mexico Loop Current and associated eddies, Brazil current), storm generated currents, and internal wave generated or soliton currents.

The limited amount of available field measurements of spar VIM indicates that turbulence in the Gulf of Mexico loop and eddy currents do not influence spar VIM response. However, there is evidence from model testing that high levels of turbulence in the model basin can affect VIM response. The structure and intensity of turbulence in ocean currents and the potential impact of current turbulence on VIM remains an uncertainty for further observation and investigation.

VIM in Current and Waves

A spar could be simultaneously exposed to large waves and currents. Low-frequency wave or wind energy transverse to the current could lead to increases in VIM (Yung et al. 2004), whereas field data (OTRC/MMS Workshop 2003) suggests that waves could reduce spar VIM. A comprehensive model test for a truss spar in waves plus current (Finnigan et al. 2005) indicate that when waves are in-line with the current, VIM can be reduced, whereas for waves transverse to current, VIM is sometimes increased as compared to current alone. This will most likely be situation specific. If the mooring or riser systems are sensitive to large motions caused by VIM and the design criteria calls for combined waves and current, model tests that address this condition should be considered.

Model Testing

Model testing has been the primary tool for VIM prediction because of difficulties in obtaining full-scale response data in a timely fashion to support projects and limited numerical or analytical approaches. Industry studies indicate, however, that model tests are only able to accurately model certain effects while compromising others. From this point of view, confidence in model test results and VIM design criteria should be established through comparison with field measurement data. The reliance on model testing, the limitations of model testing, and limited validation with full-scale data should be recognized as a potential sources of uncertainty in the design process.

A sound VIM model testing practice should pay attention to the following issues:

- hydrodynamic scaling (including free surface effect);
- geometric similitude (including modeling of appurtenances);
- dynamic similitude (including mooring stiffness and damping);
- model degrees of freedom;
- current direction and profile; and
- length of response record.

Each of these is discussed in the following sections.

Hydrodynamic Scaling

Reynolds number scaling and Froude number scaling are the two relevant scaling parameters for hydrodynamic model testing of offshore structures (Sarpkaya and Isaacson 1981, Chapter 9).

The Reynolds number is defined as

$$R_e = \frac{V_c D}{\nu}$$

where R_e is the Reynolds number, V_c is the current velocity, D is the hull diameter, and ν is the kinematic viscosity of the fluid.

The Froude number is defined as

$$F_n = \frac{V_c}{\sqrt{gD}}$$

where g is the gravitational constant.

It is not practical to satisfy both the Reynolds and Froude scaling simultaneously for the model and prototype flows. For a model dimension D that is substantially smaller than prototype, either the gravity (g) needs to be significantly increased, or viscosity (ν) of the testing fluid needs to be significantly decreased. For spar hull diameters of 70 ft. to 140 ft. and design current velocities of 2 knots to 5 knots, the Reynolds number for the full-scale structures (prototype) are in a range of 20,000,000 to 100,000,000. To match such Reynolds numbers in the model basin would require that the model experience the same hydrodynamic force as that of the prototype, which is impractical.

There are currently two basic testing approaches used in the industry:

1. *Submerged horizontally mounted models.* These tests are conducted primarily in 1 degree-of-freedom (DOF), usually in the transverse mode, with some test programs incorporating 2 DOF (transverse and in-line). Until recently, this was the only way to perform supercritical Re VIM model tests, and by limiting the DOF, it is still a good way to study some VIM fundamentals. The model is usually towed to represent a uniform current; however, such tests can be performed in a flume. When testing with a horizontal submerged model, it is necessary to use mirror image of double body with a divider plate in the center. The divider plate is used to prevent flow communication across the divider plate. The divider plate plays the role of the free surface, which acts as a barrier through which flow cannot pass. There have also been unpublished attempts to emulate a shear current by using a rotating arm basin.
2. *Vertical freely floating models.* These tests are primarily done in 6 DOF, with the mooring system usually represented by a simple four-point symmetric spring mooring system in the model basin. It is difficult to restrain these models to respond in fewer DOFs; however, it is possible in some cases to set up the mass properties such that other DOF (for example, pitch and roll) are not excited. This mode of testing is required if full 3D effects are desired and especially if other modes of response (roll, pitch, heave) could participate in the VIM response. Until recently this type of testing was restricted to subcritical Reynolds numbers. However, recent model tests have demonstrated the feasibility of performing these tests in the supercritical range (Finnigan 2007). For such tests, care must be exercised to keep the Froude number less than approximately 0.25 to minimize the damping effect bow waves could induce in the response. However, there are only a few basins in the world capable of performing VIM tests of the large models required.

Geometric Similitude

The geometry of the hull, strakes, and appurtenances should be accurately scaled. This includes construction openings in the strakes, support brackets, chains, anodes, external pipes, and other structural elements that may affect the flows around the body. Accurate modeling of appurtenances is particularly important in developing VIM directional sensitivity and testing effectiveness of VIM suppression devices such as strakes.

Some members, such as the truss members of a truss spar, may result in viscous damping effects that are Reynolds number dependent. These members should be sized to result in an appropriate amount of damping in model tests. Sometimes the truss members are removed or reduced significantly in size in the model tests to avoid issues about Reynolds scaling and damping that they may cause. This would most likely result in a conservative estimate of the VIM amplitudes. If this is done, care must be taken to ensure that the dynamic similitude of the model is maintained (see next section).

Dynamic Similitude

Dynamic similitude requires that the rigid body dynamics of the full-scale and model-scale systems be similar. Dynamic scaling is associated with the vessel's rigid body modes, mass ratio, and reduced velocity. However, modeling all of the rigid-body modes may not always be important. If a spar might lock-in to sway at lower velocities and lock-in to roll at higher velocities, the 2 DOF might actually couple or lock-in simultaneously (Finn et al. 2003). In this case it is important that the sway and roll modes and periods be properly scaled. If transverse sway is the dominant VIM response, and the roll period is much less than half of the sway period, then tests with a 1- to 2-DOF rigid body may be sufficient (Sandström et al. 2003). Caution should be employed if testing in 1 DOF, because restraining the surge degree-of-freedom can lead to underestimation of the sway response (Jauvtis and Williamson 2004).

The mass ratio can also have a large effect on the range of lock-in, as well as the amplitude (Feng 1968, Sarpkaya 2004, Williamson et al. 2004). The mass ratio for a free floating body is by definition equal to 1.0 (displacement = weight). This mass ratio should be maintained for the model tests.

The mooring stiffness is an important parameter, but depending upon the project needs it may not be necessary to precisely model the prototype stiffness. There are two approaches to modeling the stiffness distribution of the prototype mooring system:

1. *Use the reduced velocity (V_{rn}) as a design parameter.* In this case, the VIM response in the model is related to the design a/D via the reduced velocity. In the model tests, the spar response is measured at different reduced velocities. In the design phase, the transverse period of the spar (hence the V_{rn}) is calculated at different offsets. At each offset, the a/D is based on the V_{rn} at that location. In this approach, a linear symmetric mooring system can be used for the model test set-up.
2. *Model the actual spread mooring of the spar.* In this case the current speed is the design parameter rather than the reduced velocity. Each mooring line or group of mooring lines is modeled by an equivalent model mooring line. The horizontal force-displacement characteristic of each mooring line or group is modeled by a bi- or tri-linear spring system so as to mimic the nonlinear force-displacement characteristic of each mooring line or group. This allows for modeling of the complete nonlinearity and asymmetry of the stiffness. For some mooring systems, such as a grouped mooring system, the asymmetry may contribute to highly directional VIM response.

Damping can affect VIM response; therefore, the damping (hydrodynamic and mechanical) generated in the model basin should be consistent with the damping expected in the field. Because mechanical damping may be generated by the testing equipment and is absent in the field, care must be taken to understand the effect of damping on the VIM response and to mitigate such effects (van Dijk et al. 2003). Hydrodynamic damping in the model test due to mooring lines and wave effects should be given careful consideration when estimating the amplitude of full-scale VIM.

Model Degrees of Freedom

Models of single degree and multiple degrees of freedom have been used. For the single-degree-of-freedom model, which is primarily used in supercritical Reynolds number testing, only transverse VIM is allowed. Two-degree-of-freedom models have been used in both sub- and supercritical Reynolds number testing. For these tests, both transverse and inline VIM is allowed. Until very recently (Finnigan et al, 2007), multiple degrees of freedom models have only been used in subcritical Reynolds number tests. These models are free to respond in all 6 DOF. The relative importance of the multiple degrees of freedom model is determined by the level of structural and hydrodynamic coupling between motions of different degrees of freedom. It is possible, but difficult, to restrain some of the degrees of freedom for a freely floating body.

Current Direction and Profile

VIM response for a spar is sensitive to small changes in current directions. Fine heading resolutions (for example, at 5°- to 10°- increments) may be required to capture the maximum VIM response.

Tow tests simulate a slab current uniform with depth. In practice, currents have a profile with current speeds that generally decrease with depth. It is possible to approximate shear current profiles in tow, flume, and basin tests (Finnigan et al. 2005, van Dijk et al. 2003); however, attempts to generate shear current profiles in model tests often generate excessive turbulence. Measurements in the loop current (unpublished) indicate that the turbulence is relatively low. Hence, careful consideration needs to be given while interpreting VIM responses in the presence of turbulent flow. Turbulence in laboratory generated shear flow can be mitigated by using varying density/viscosity stratified liquid layers in the model tests or by immersing the spar in a false hole in the floor so that only the upper portion is exposed to current while the lower portion is in quiescent flow (Finnigan et al. 2005). For this second option, care must be taken to minimize the turbulent boundary layer that is introduced by the false bottom.

Length of Response Record

Sufficiently long response time histories are required to provide meaningful statistics such as standard deviation, significant, and maximum values. The length of time history required depends on the periodicity of the VIM response (Sandström et al. 2003). When the VIM motion is well developed and sustained (for example, fully locked-in), relatively few cycles are required to establish the maximum VIM amplitude. If the VIM response is modulated (for example, in the lock-in and lock-out regions), long records are required to establish reliable statistical values. For highly modulated response, more than 100 cycles may be required, or alternatively, several repeat tests may be performed (Finnigan et al. 2005). The startup transient response should be excluded from the records for statistical analysis.

Concluding Remarks

Most model tests conducted in support of spar projects today are done with a vertically moored spar with lower than super-critical Reynolds number conditions. These tests are typically conducted by towing the model with four mooring springs in a large towing basin. Limited tests have been conducted using a horizontal, submerged model in high-speed towing tanks (OTRC/MMS Workshop 2003; Yung et al. 2004; Sandström et al. 2003), and tests (both vertically and horizontally moored) in a circulating flume (unpublished). Recently, tests have been conducted with a vertically moored spar under supercritical Reynolds numbers (Finnigan et al. 2005).

Model tests have not been performed for all spars. VIM response itself is self-limiting. If a bounding analysis indicates that the mooring or riser systems will not be governed by high current or VIM responses then VIM tests may not have to be performed (Magee et al. 2003).

References

Edwards, R., Shilling, R. Theti, R. and Karayaka, M., "BP Horn Mountain Spar—Results of Comprehensive Monitoring of Platform and Riser Responses," Deep Offshore Technology Conference and Exhibition, Marseille, France, 2003.

Feng, C. C., "The Measurement of Vortex-Induced Effects on Flow Past Stationary and Oscillating Circular and D-Section Cylinders," M.A. Sc. Thesis, University of British Columbia, Canada, 1968.

Finn, L.D., Maher, J.V., and Gupta, H., "The Cell Spar and Vortex Induced Vibrations," OTC 15244, 2003.

Finnigan, T., Irani, M, and van Dijk, R, "Truss Spar VIM in Waves and Currents," OMAE 2005-67054.

Finnigan, T., Roddier, D., "Spar VIM Model Tests at Supercritical Reynolds Numbers," OMAE 2007-29160

Halkyard, Sirnivas, Constantinides, Oakley, and Thiagarajan. "Benchmarking of Truss Spar Vortex Induced Motions Derived from CFD with Experiments," Proc. of the 24th Int. OMAE Conference, OMAE2005-67252, 2005.

Jauvtis, N. and Williamson, C. H. K., "The Effects of Two Degrees of Freedom on Vortex-Induced Vibration at Low Mass and Damping," *J. Fluid Mech.,* Vol, 509, pp 23-62, 2004.

Magee, Allan, Anil Sablok, Joe Gebara, "Mooring Design for Directional Spar Hull VIV," Proceedings, OTC 15243, 2003.

Oakley, O., Constantinides, Y., "CFD Truss Spar Hull Benchmarking Study," OMAE 2007-29150.

OTRC/MMS Workshop, "Spar Vortex-Induced Motions," Proceedings of Workshop, Navasota, Texas, October 22-24, 2003.

Sandström, R. E., Yung, T-W., Slocum, S. T., and Ding, Z. J., "Advances in Prediction of VIV for Spar Hulls," Deep Offshore Technology Conference, Marseille, France, 19-21 Nov 2003.

Sarpkaya, T. "A Critical Review of the Intrinsic Nature of Vortex-Induced Vibration," Journal of Fluids and Structures 19 389–447, 2004.

Sarpkaya, T. and Isaacson, M., "Mechanics of Wave Forces on Offshore Structures," Van Nostrand Reinhold Company, New York, NY, USA, 1981.

Skip, R.A., Griffin, O.M. and Ramberg, S.E. "Strumming Predictions for the SAECON II Experimental Mooring," Offshore Technology Conference, OTC 2491, 1977.

van Dijk R. T., Voogt A., Fourchy P. and Saadat M., "The Effect of Mooring System and Sheared Currents on Vortex Induced Motions of Truss Spars," Proceedings 22nd International Conference on Offshore Mechanics and Arctic Engineering, OMAE'03, Cancun, Mexico, 2003.

Vandiver, J.K., "Drag Coefficients for Long-Flexible Cylinders," Offshore Technology Conference, OTC 4490, 1983.

Williamson, C.H.K. and Govardham R., "Vortex Induced Vibrations," Annual Review of Fluid Mechanics, 36:16.1-16.42, 2004.

Yung, T. W., Sandström, R. E., Slocum, S. T., Ding, J. Z., and Lokken, R. T., "Advancement of Spar VIV Prediction ," OTC 16343, 2004.

Chapter 8: Deepwater Spar Model Testing: Considerations for Planning a Physical Model Test Program

By Peter Johnson, Project Manager, BMT-Scientific Marine Services, Inc.

Abstract

Considerations for planning a model test program for a deepwater spar platform include model scaling, the capabilities of the test facility, the environments to be tested, the test duration, and the responses to be studied. Spar platforms are presently being designed and installed in water depths that exceed the capability of even the largest model basins to replicate at reasonable scales without distorting or truncating the model mooring system. Numerical models are called upon to extrapolate experimental results to full scale. The relative motion between top tension risers, the spar hull, and the friction force that arises from contact between these components has been observed to be important to the global motions. Vortex shedding is typically studied by towing models to minimize the ambient turbulence. The effect of large Reynolds numbers, like ones typically found for spars in ocean currents upon vortex induced motion, has not found a consensus among designers or experimentalists.

Introduction

Deepwater spar platforms have some unique considerations when planning a physical model test program. Care must be taken in the selection of the model scale, design of the experiments, design of the model, and selection of the instruments to capture the phenomena of interest. The particular capabilities of the test facility—size, environments to be generated, and measurement capabilities—must be taken in account.

Examples of scale models of the two most common types of spar platforms for offshore oil production are illustrated in Figures 8-1 and 8-2. The shell of the classic spar, shown in Figure 8-1, is a circular cylinder through the full depth of the hull. Enclosed within the hull is a large quantity of water, giving the platform a large mass with long natural periods of motion relative to typical ocean wave periods. The example of a truss spar, shown in Figure 8-2, replaces the lower portion of the hull with a truss structure. The truss includes several horizontal plates to capture the water mass and also results in long period motions similar to the classic spar.

Fig. 8-1. Example of classic spar

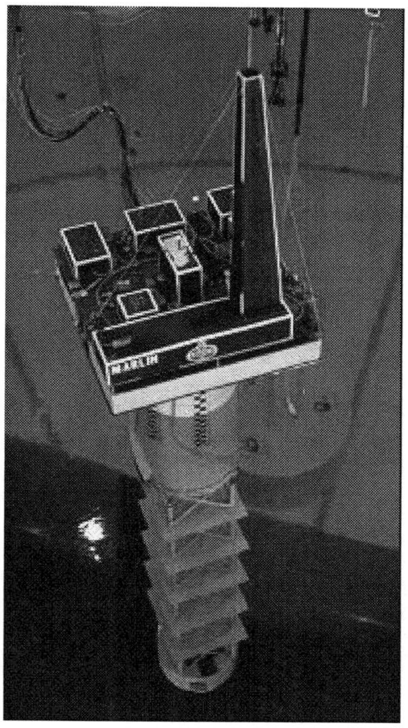

Fig. 8-2. Example of truss spar

Physical Model Testing Objectives

The objectives of deepwater spar model test programs can be varied and complex. Multiple objectives can result in model design requirements that may be contradictory or difficult to resolve.

Tests may be conducted to study different phases in the life of a platform, including:

- dry transport;
- wet transport;
- installation; and
- in-place performance.

Dry transport is usually for long distances on a heavy lift vessel or barge from the fabrication yard to a location near the site of operation. Issues that may be addressed with physical modeling include quantification of tie-down loads, local slamming forces, and global motions.

For wet transport to the installation site, the spar hull is floated horizontally, temporarily exposing it to wave and current loads that it will not experience during in-place operation. In wet transport experiments, local forces due to waves, wave run-up, structural loads, and global motions may be the subject of physical model experiments.

Installation on site includes upending the hull from the horizontal to the vertical. The upending process may be modeled to study motions and hull structural loads.

In-place model test programs maybe very extensive and typically address a variety of subjects. Test environments may include extreme events, such as hurricanes and loop currents, as well as normal operating conditions to address fatigue. Examples of in-place spar model test objectives include:

- Calibrate and/or validate numerical models
- Quantify mean, RMS and extreme responses
  a. Global motions
  b. Wave effects
  c. Structural loads
  d. Mooring loads
  e. Riser loads
- Quantify vortex induced vibration (VIV) response

Model Scale Considerations

Scaling Laws

Wave models are typically scaled using the Froude scaling law:

$$Fr = \frac{U_p}{\sqrt{gL_p}} = \frac{U_m}{\sqrt{gL_m}}$$

where
Fr = Froude number
U = characteristic velocity
g = gravity
L = characteristic length
Subscript p refers to the prototype
Subscript m refers to the model

Physical models of spars in waves attempt to preserve the Froude number, resulting in the scale relationships listed in Table 8-1.

Table 8-1. Froude Scale Relationships

Quantity	Scale Relationship
Length	λ
Area	λ^2
Volume	λ^3
Time	$\lambda^{1/2}$
Velocity	$\lambda^{1/2}$
Acceleration	1.0
Force	$(\rho_s/\rho_f)\,\lambda^3$
Moment	$(\rho_s/\rho_f)\,\lambda^4$
Pressure	$(\rho_s/\rho_f)\,\lambda$

where
$\lambda = L_p/\,L_m$
ρ_s/ρ_f = saltwater density to freshwater density

The selection of a model scale for a test program invariably results in compromises. Competing with the inertial and gravity effects on a platform are the viscous effects of the flow around the platform and the mooring. Viscous modeling is scaled by the Reynolds scaling law:

$$Re = \frac{U_p L_p}{\nu} = \frac{U_m L_m}{\nu}$$

where
Re = Reynolds number
ν = kinematic viscosity of water

Froude scaled spar models have a Reynolds number that is too small, resulting in viscous damping effects that may be too large. Care must be taken when applying damping values derived from scale model tests to full scale.

Test Facility Capabilities

Many recent deepwater platforms are located in water depths that exceed the ability to replicate acceptably using even the largest model basins.

The scale limits for defining small models varies with test facility and is based upon the capabilities of the basin in question. Modeling limits may be defined by the basin wave generators and wave absorbers, which are designed to operate within a range of model wave periods and wave heights. Similarly, current and wind generators are designed to provide flow fields with a limited range of velocities, which may be uniform and homogeneous over only a limited test volume. As models become smaller the instruments used to make measurements may increasingly distort the responses under study. Finally, the precision of the measurements may be degraded as the absolute magnitude of the responses becomes smaller for smaller models.

Environments

Wave Generation

Planning for a physical model test must consider the capabilities of the test facility to create the test environments. Care must be taken, especially when the test program includes both survival and operational sea states, so that the waves can be made over the full range of periods of interest.

The type of waves to be generated will depend upon the platform design criteria and the calibration and verification requirements of the numerical models used in the platform design. Spar model test programs may include regular waves, random waves, or both. Survival and operational sea states may be of mutual interest.

All wave generators are designed to perform within limits. Figure 8-3 is an example of the regular wave performance envelope for a typical wavemaker. Waves, both in nature and in the laboratory, have a limiting steepness at which they begin to break. This limit is indicated in Figure 8-3. Attempting to make waves beyond this natural limit will only result in waves breaking directly off the waveboards and will not increase the wave height at the model. The second wave height limit is a mechanical one imposed by the maximum displacement of the waveboards. For increasingly longer wave periods, the wave height is becomes less for constant waveboard displacement (stroke) as is also indicated in Figure 8-3.

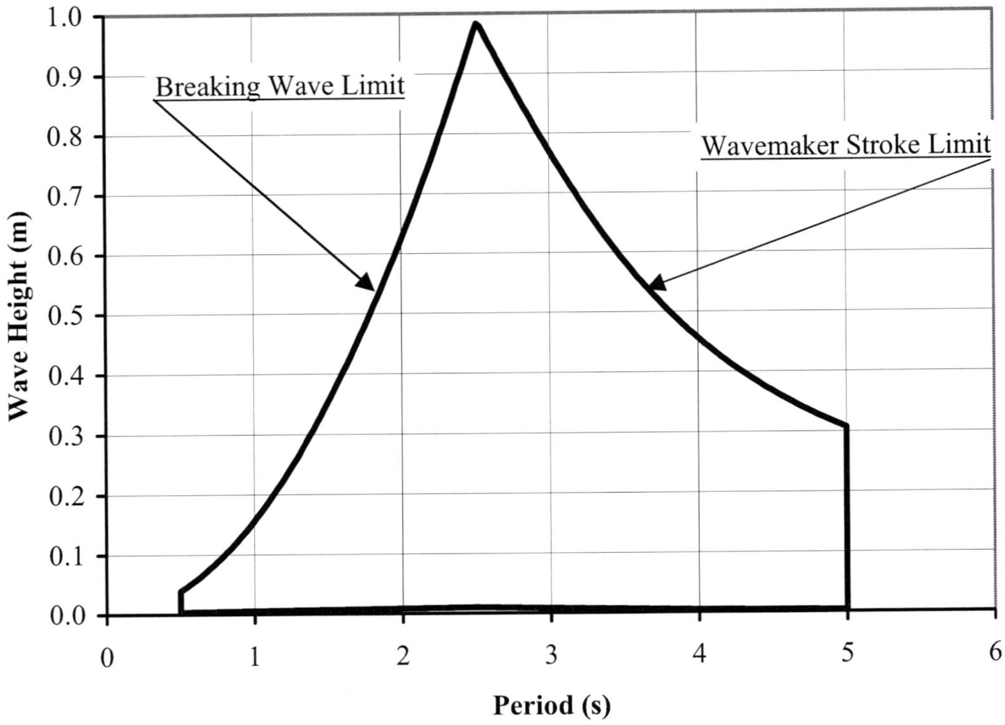

Fig. 8-3. Example of a wave generator performance envelope

The conditions for random wave tests frequently specify that they be performed in conjunction with wind and current. It has been observed that model current may heavily modify the waves (Phillips 1977). Similarly, model waves may modify the current profile. When calibrating combined waves and current, it is important to do them together. To separate the effect of the waves and the current on the spar, it is advisable to perform wave alone tests as well.

Wave Statistics

Typically, model sea states are created and calibrated prior to installation of the model platform in the basin. They are usually specified in terms of a spectral shape, significant wave height and spectral peak period. In many test programs multiple realizations of a sea state are made, that is, the same spectral properties are used to generate different wave trains using different random "seeds." In test programs with multiple realizations a choice should be made to require either each wavetrain closely matches the target spectrum or only the average of all the realizations closely matches the target.

In addition to specifying the sea state's spectral parameters, it has become common practice to closely examine the distribution of extreme wave height and extreme crest elevation because of the consequences for airgap, wave overtopping, and deck elevation. When generating random waves it is important to remember that the largest crest height in an experiment is a random event with an associated probability distribution and that the largest single event will be different for different model wave trains derived from the same spectrum. As an example, consider that the expected largest crest elevation in 1000 waves that follow the Rayleigh distribution is approximately 0.93 times the significant wave height (H_s). However, the actual crest elevation in a randomly selected model wave train, with the specified H_s, will probably only fall within some tolerance of the expected value. This tolerance has an associated level of probability. For instance, the 95 percent confidence interval for the one in 1000 crest elevation in a Rayleigh distribution is $(0.93 \pm 0.085)H_s$. Narrow tolerances on extreme values, such as crest elevation, may result in rejecting statistically valid model wave trains. If extreme values are important to the test program's objectives, longer test durations or multiple realizations of the same sea state are required to obtain a distribution of the extreme values that is stable at the specified probability or return interval.

Wind

Model wind, or an equivalent wind force, may be imposed on a spar model to obtain surge offsets and trim angles similar to those imposed on the prototype from real wind. The specified target wind force values are usually developed from wind tunnel tests or numerical wind models.

In a wave basin, model wind is typically imposed using fans. In this case, the wind speed and/or the model deck sail area are adjusted to obtain the specified forces, offsets, or rotations. In some model basins, the low frequency variations in the wind speed may be included in the experiments by changing the speed of the fans over the duration of the test. Calibration of the wind force is usually accomplished in tests with wind alone. The model wind spectrum may be made to match a specified target spectrum, much as the waves are required to match a specified sea state spectrum.

Model wind has the advantage compared to the imposition of an equivalent force: as the model moves, the instantaneous wind force also changes due to the change in projected area. The disadvantage of model wind is that it is typically calibrated and valid over a limited region to which the platform is restricted for valid results. Wind calibrations usually require multiple measurements at several locations both horizontally and vertically to adequately describe the vertical wind profile and ensure that it is homogeneous over the volume occupied by the model during the experiments.

Another disadvantage is that tests with model wind are less repeatable than those in which a wind force is imposed. Usually the airflow is unconstrained by guide walls so uncontrolled, temporary fluctuations occur even though the mean speed, standard deviation, and low frequency spectrum may repeat well from test to test.

If model wind is not employed, frequently an equivalent wind force will be imposed at the specified center of wind pressure. Typically, the equivalent wind force is imposed by pulling on the spar model with a lightweight string, using a tension-measuring instrument to record the force. Low frequency wind force variations may be imposed using variable force devices.

An equivalent wind force has the advantage that the force can be measured and is precisely known at all times during an experiment. It has the disadvantage that, as the spar pitches and yaws it changes the projected area exposed to the wind. In this case, the imposed model force may not be representative of the prototype. A further disadvantage is that the model wind force may include unwanted wave frequency components due to the motion response of the platform. Typically, this is less of a problem with spars, which have relatively small motions at wave frequencies, than it is with smaller platforms whose motions may be dominated by wave frequency forces.

Current

Model current, or an equivalent current force, may be imposed on a spar model to obtain offsets that would occur in the prototype due to real current.

Current calibrations are typically done at several horizontal and vertical locations to ensure that the vertical profile matches the specification and that it is relatively homogeneous over the volume occupied by the model during the experiments. Model currents and waves affect each other, and waves should be calibrated in the presence of the current. Similarly, model currents should be calibrated in the presence of the waves.

It is difficult to generate a current flow field without introducing turbulence. Data analysts should recognize that experiments with model current are inherently less repeatable due to uncontrolled fluctuations in the flow.

Large amounts of ambient turbulence in the flow field may corrupt observations of vortex induced vibration (VIV). VIV is typically studied by towing the model in a calm basin. The platform mass and the mooring stiffness should be represented if the periods of oscillation and the amplitudes are to be representative of the prototype. Several sway cycles should be observed in an experiment, so a relatively long towing length is required to capture the long period motions.

Test Duration and Multiple Realizations

Since spars typically have long natural periods of motion, long duration experiments are necessary to observe a sufficient number of cycles to obtain statistically stable results. Common practice is to require the full-scale equivalent of 3 hours of experiments so that the number of low frequency cycles for motions, such as surge or sway, are in the range of 40 to 100.

The start of an experiment with waves, wind, or current imposes a transient response on a spar model as it goes from a calm environment to a dynamic one. This initial transient response should not be included in the analysis if the platform's response to a stationary environment is to be quantified.

If extreme values are important then multiple intervals of the return period of interest should be run. Frequently, the design return interval for an extreme value is three hours at full scale. In this case, if the test duration is three hours only one extreme value is obtained. Multiple realizations (several experiments) should be performed to obtain a distribution of extremes and a stable estimate at the probability level of interest.

The length of time that an individual experiment may be run without wave reflections corrupting the experimental results must be considered. Therefore, the performance of the beach or wave absorber is important. Furthermore, reflections from the model may be re-reflected off the wavemaker and the basin walls causing the tested wave environment to deviate from the calibrated environment. Care must be taken to ensure that the model wave environment the platform is exposed to is similar to the environment that was calibrated by limiting the test duration.

Transportation Tests

Typically, spar platforms are constructed in shipyards remote from their site of operation. Spar platforms may experience long-distance, trans-oceanic shipment from the fabrication yard and shorter wet tows from a final assembly area to the site of operation. Transportation model studies may be performed to validate numerical models or to quantify temporary loading conditions, such as direct wave impacts on heave mass plates.

Installation Tests

Typically, spar platforms arrive at the site with the hull floating on its side. The hull must be upended from the horizontal orientation to the vertical orientation for installation and operation. This is typically done by adding ballast to tanks located at the keel of the spar. Upending may be done rapidly by allowing the keel tanks to freely flood and sink under their own weight or more slowly by lowering the keel in a controlled manner using a winch.

Physical model tests of the upending process have been used to study

- the motion of the platform during upending; and
- the structural stresses at the joint between the hard tank and the lower hull.

In-Place Tests

Motions

Spar natural period motions are typically much longer than ocean wave periods. As previously noted, the initial transient motion decay interval at the start of an experiment should not be included in the analysis of an experiment that should posses a stationary environment and response. The test should also be of sufficient duration or include multiple realizations to ensure that sufficient low-frequency oscillations are captured for stable statistics. Approximately 100 cycles are typically sufficient to quantify mean and rms response to obtain stable statistics. For extreme value estimates, many more cycles may be required.

Vertical motions may have small hydrodynamic damping, particularly for classic spars. Care should be taken to ensure that other damping forces, such as arise from riser to guide friction, are accurately measured and recorded.

Structural Loads

Bending moments may be important in the design of a spar. Frequently, model bending moments are measured at the connection of the hard tank to the lower portion of the spar. It is usually important to calibrate the bending moment transducers in place in the hull to quantify any structural cross-talk effects.

Frequently, the vertical loads on the mass plates of a truss spar are measured in model tests. The amount of mass trapped between the plates and the amount of damping obtained due to flow around the plates translate into structural loads. These loads may be highly repetitive, raising the potential for structural fatigue.

Occasionally, local loads may be measured on specially designed elements of the model, such as under-deck girders, wave deflectors, and strakes.

Runup and Airgap

Waves in the immediate vicinity of the hull may be difficult to model numerically with high precision, particularly due to the non-linearities in the wave crests. Physical models provide a good way to study issues such as deck elevation.

Since it is not possible to measure the water surface elevation everywhere under the deck of the model, point measurements from wave probes may not be the most reliable means of characterizing the airgap. Frequently, the best method to determine if a wave has hit the deck is to review the videotape of the experiment. For the videotape to be informative, the camera must be placed where it can obtain a view level with the bottom of the deck. Waves passing under the deck will be obscured if they are viewed from too high an elevation.

Centerwell Pumping

Many spar platforms have a centerwell for passing the top tension risers from the deck through the keel. The water in this centerwell may move in response to the waves and the motions of the hull. Resonant motion of the water surface has been observed both pumping vertically and sloshing horizontally, especially in truss spars. The amount of motion in the moonpool depends on the depth to opening to the sea and the number and size of the openings at the bottom.

Mooring

As previously noted, many spar platforms are located in water that is too deep with the seafloor area covered by a mooring that is too large to model reliably in even the largest model basins. Deepwater spar mooring models are usually designed to produce only limited similitude with the prototype. Numerical models are called upon to extrapolate experimental results to full scale.

The standard modeling practice with physical models of shallow-water moorings has been to reproduce most of the elements of each individual mooring line with as little distortion as possible. In these cases, full-scale wire rope is represented by small cable and full-scale chain is represented by jewelry chain. The mass distribution is carefully preserved to achieve the proper catenary shape of the model mooring elements. The scaled axial stiffness of the lines is obtained with in-line springs. Anchor locations and mooring line touchdown points are precisely modeled. This careful attention to detail ensures that the static characteristics of the model mooring will accurately represent the prototype while the dynamic characteristics will be a reasonable approximation.

Deepwater spar mooring models cannot preserve the anchor locations, line lengths, or mass distribution of the prototype moorings. Because so much of the mooring line is not physically represented, the line shape, motions, and damping characteristics are poorly represented by the model. Model deepwater moorings are typically designed to preserve the mean vertical force and the static offset force vs. displacement (surge or sway) of the prototype. If possible, mooring system overturning moments may be preserved. Larger diameter model elements may be used to attract additional viscous damping to make up for some of the damping from the truncated portions of the lines. In-line springs are usually used to adjust the overall stiffness of the mooring system. Nevertheless, line dynamics are not well replicated at scale in truncated or distorted model mooring lines (Chen 2000).

The model mooring should be well documented so that a numerical model can be developed. Anchor locations, line lengths, wet weights, and dry weights should be measured and recorded for every individual element of each model line. Care should be taken to ensure that the physical model is accurately represented in the numerical model. Piecewise linear springs and clump weights may be difficult to include in numerical mooring models and should be avoided if possible. Long, heavy model elements that drag across the basin floor may introduce friction and hysteresis into the static offset curve and should be noted.

Numerical models should be calibrated and validated by modeling the properties of the model spar and the truncated model mooring. Static offset tests and free decay tests are important for calibrating numerical models of the physical model. Forced oscillation tests may be necessary to obtain model scale added mass and damping coefficients at frequencies other than the natural frequency. The test program should be designed to separate, as much as possible, the effects of waves, wind, and current. This may be accomplished with wave alone, wind alone, and current alone tests. However, care should be taken in the analysis since model waves and currents interact in a way that is not simply additive.

Extrapolation of tests results from truncated model moorings to full scale should be done using numerical models. Numerical models that accurately reproduce the statics and dynamics of the scaled physical model can be applied to the full situation with care to adjust for full scale mooring line dynamics.

Risers

Top Tension Risers

Top tension risers may significantly affect spar motions through the side loads they impose at the keel and through the Coulomb friction force that arises due to the relative vertical motion of the hull and risers (Prislin et al. 1999). Modeling the relative motion of the risers is also important for determining the limits of motion for design of the tensioners, flowlines, and umbilicals.

The friction factor between the components of the riser and the spar hull is modeled by selecting materials similar to those used in the prototype. The normal force is obtained by ensuring the Froude scaled buoyancy, weight, and riser tension are obtained.

Usually, all the top tension risers involved in a specified scenario are modeled. The riser tensioning scheme is also usually modeled at some level of detail. Some spars provide tension to the risers using floating buoyancy cans located inside the hull. These buoyancy cans may move vertically with respect to the hull as well as horizontally within the restriction of their guides. The friction factor between the buoyancy cans and the guides, as well as the riser keep joint and the keel guide, should be measured and modeled.

Some spars employ deck-mounted riser tensioners similar to those on other types of offshore production platforms. If there is significant friction in these tensioners then it may be necessary to develop model tensioners that reproduce this friction force.

Truncated risers impose distorted side loads on the spar keel. This may affect both the pitch motion of the model and the heave motion through distortion of the friction force.

Steel Catenary Risers

Steel catenary risers typically add a small, nearly constant offset force and trim moment to spar platforms. In the case of truncated model mooring systems, it is usually not possible to include the change in riser tension due to displacement of the platform because it is very small compared to the effect of the mooring. The effect of the vertical component of the SCR load is frequently incorporated into the model ballast and the horizontal load is imposed using a soft horizontal spring attached at the SCR porch. Alternatively, the SCR may be represented by a soft spring with a representative vertical angle.

VIV Tests

Experiments to examine VIV in spar platforms are usually performed by towing the model.

Vortex shedding may be characterized by the Strouhal number, defined as:

$$S = \frac{fL}{U}$$

where
S = Strouhal number
f = vortex shedding frequency
L = a characteristic length
U = flow velocity

Since vortex shedding is dominated by the geometry of the spar and the fluid viscosity, the Reynolds number should be considered. Some experimenters observe that vortex shedding is a weak function of the Reynolds number and seek only to obtain a value beyond which the Strouhal number does not change significantly. Other experimenters consider that this observation is not proven for Reynolds numbers that are representative of spars in ocean currents on the order of 10^6 or greater.

Towing tests have been done to assist in the design and assess the effectiveness of vortex suppression devices such as strakes. These tests should be conducted with a spar model that is free to sway since vortex shedding is affected by the motion of the platform. The tests should be of sufficient length to permit multiple oscillations.

Froude scale (low Reynolds number) towing tests may be conducted in typical towing tanks at conventional model towing speeds. High Reynolds number towing tests are typically conducted at very high speeds and require very specialized test fixtures to accommodate the large model forces. Relatively few towing tanks have the length and speed capabilities to conduct high Reynolds number spar VIV tests.

Conclusions

Planning for model tests of deepwater spar platforms requires careful consideration of the uses the results will be applied to. Model waves, winds, and currents should be designed taking into account typical characteristics of spars, such as long periods and lightly damped motions. Special care should be taken to design experiments that account for the lack of similitude in the model moorings and risers. Numerical models should be used in conjunction with the physical models to apply the experimental results to full scale. Consensus is lacking among designers and researches about the proper scaling of vortex shedding on large structures such as spars.

Appendix: Editor's Comment on New Hybrid Model Testing

Mooring lines and risers are expected to play a larger role for deepwater and ultra-deepwater development. Thus, it is necessary to test the system with the full length of lines. No wave basin in the world is able to test such a deepwater platform with the right scale mainly due to the depth and width limitation. Therefore, role of numerical modeling, which can properly model vessel motions and riser-mooring dynamics and their interactions, becomes more important. Most researchers and engineers agree that the experimental-numerical hybrid modeling testing should be a circumventive solution for deepwater and ultra-deepwater projects. In other words, first test the system in the basin with equivalent truncated mooring-riser system, use the experimental data to calibrate the reliable numerical model, and then use the numerical model to analyze the full-depth system. For successful hybrid model testing, very reliable numerical models should be developed and verified.

References

Chen, X., Zhang, J., Johnson, P., and Irani, M. (2002). *Studies on the Dynamics of Truncated Mooring Lines*, Proc. 10th Intl. Offshore and Polar Engineering Conf., Seattle, Vol. II, pp. 94-101, June 2000.

Phillips, O.M., (1977). The Dynamics of the Upper Ocean, Cambridge University Press, Cambridge, 2nd Edition.

Prislin, I., Halkyard, J., DeBord, F., Collins, J.I., Lewis, J.M., *Full Scale Measurements of the Oryx Neptune Production Spar Platform Response*, OTC 10952, Offshore Technology Conference, Vol. II, May 1999.

Chapter 9: Comprehensive Full-Scale Data Comparison for the Horn Mountain Spar

By Arcandra Tahar, FloaTECH, Houston, Tex.; John Halkyard and Lyle Finn, Technip, Houston, Tex.; and Pierre Liagre, BP, Houston, Tex.

Abstract

The Horn Mountain production spar, installed in 1654 m of water in the Gulf of Mexico in June 2002, was the deepest floating production unit in the world at that time. A comprehensive instrumentation program was initiated to measure spar and riser responses (Edwards et al., DOT 2003). Three Offshore Mechanics and Arctic Engineering (OMAE) papers (Halkyard et al., OMAE 2004, Tahar et el, and OMAE 2005, 2006) reported the correlation results.

This chapter discusses these measurements and compares them with analytical predictions for spar behavior for two selected events, Hurricane Isidore in September 2002 and a summer storm in August 2003. Analysis focused particularly on the slowly varying surge and pitch motions, the importance of coupling with risers and mooring on hull motions, the importance of Coulomb friction between wire chain and the fairlead bearing to the dynamic tension of mooring lines, comparisons of various numerical methods, and the effect of directionality of environmental conditions.

Introduction

Since the installation of the first spar, Oryx Neptune, in the Gulf of Mexico in 1996, spars have proven to be an economic and reliable solution for oil production in deep and ultra-deep water. The spar is particularly well suited for remote areas of the Gulf of Mexico. To date, three spar concepts have been proposed:

- classic spars, which use a deep-draft hollow vertical cylinder;
- truss spars, which use a combination of relatively shallow-draft hollow cylinder and truss structure extended to a soft tank. (Figure 9-1 shows a typical truss spar with mooring lines and risers); and
- cell spars, which use a combination of small-diameter tubes (Finn et al. 2003).

As of March 2007, three classic and 11 truss spars are in production, and two truss spars are under construction.

For floating systems like the spar, the hull, mooring lines, and risers constitute highly interdependent and highly nonlinear systems. In theory, as water depth increases, the ratio of the mooring lines and risers mass to the hull mass increases. Inertia and damping effects are expected to be greater as well. Low-frequency pitch and surge motions are in turn reduced, which should improve the loads and fatigue life of the risers.

The spar is different from the tension leg platform (TLP). Because of the ringing or springing on a TLP, coupling effects can in fact magnify the hull's extreme responses. As a result, hull/mooring/riser coupled dynamic analyses need to be employed.

Four methods are commonly used to analyze motions:

1. Using quasi-static coupled analysis, the mooring and risers are modeled as massless, linear, or nonlinear springs. Hull responses can be calculated based on estimated mooring and riser reactions derived from tensions produced from a static-offset relationship.
2. A semi-coupled dynamic analysis is performed where dynamic responses for moorings and risers are computed based on the motions at the attachment points based on quasi-static coupling.
3. Hull, mooring, and riser responses are solved simultaneously as a dynamically coupled system in the time domain.
4. Hull, mooring, and riser responses are solved simultaneously as a dynamically coupled system in the frequency domain.

This study uses two simulation programs. The first computer program, called MLTSIM, uses the coupled quasi-static analysis method, hydrodynamic code for time domain simulation of offshore structures (Glanville et al. 1991). The program allows modeling of diffraction, radiation damping, and wave drift forces on panelized bodies in addition to loads on Morison elements.

The second computer program called WINPOST falls into the coupled analysis method category. This is a time and frequency domain program for the hull/mooring/riser coupled static/dynamic analysis of floating offshore structures (Ran and Kim 1997, Kim et al. 1999).

WINPOST uses the elastic rod technique to model the mooring system. It is ideal for small strain, large displacement structural analysis of slender members such as tethers, risers, and catenary mooring lines. This technique uses a single global coordinate system in the finite element formulation of the rod model. Therefore, the model is simpler and more efficient than other conventional nonlinear models, such as the updated Lagrangian beam model.

Fig. 9-1. Typical truss spar with mooring and risers

Specification of Horn Mountain Spar

The Horn Mountain development is in 1654 m of water, approximately 100 miles (160.9 km) southeast of New Orleans in the Gulf of Mexico. This deepwater field, operated by BP, lies in Mississippi Canyon blocks 126 and 127.

Total operating topsides weight for the spar is 8900 tons. The spar hard tank is 32.3 m in diameter with a bottom depth of 53.64 m, a length of 169.2 m, and a draft of 153.9 m. A 15.8 by 15.8 m center well provides 4 by 4 slots at 3.96-m spacing to accommodate up to 14 top tension risers (TTRs)—buoyancy can.

The mooring system features three groups of three semi-taut, chain-wire-chain mooring lines. The fairleads are 3.7 m above the bottom of the hard tank, which places them 50 m below the mean water line. Figures 9-2 and 9-3 show the mooring configurations and hull, respectively, while Tables 9-1 and 9-2 tabulate the properties of the as-built mooring lines.

Table 9-3 shows an estimate of TTR and SCR (steel catenary risers) properties during the two events (Hurricane Isidore and a summer storm).

According to the operations log, the position of the spar is 8.2 m off of the nominal centerline before the hurricane. Hence, pretensions in Table 9-2 are different from those in effect during the two. The difference can have an effect on the tension comparison between simulation and field measurement.

Table 9-1. Mooring Line Properties

Line Properties	Unit	Platform Chain	Riser Wire	Ground Chain
Type		R4 Studless	Spiral Strand	R4 Studless
Diameter	mm	146.1	128.0	146.1
Length				
Group 1	m	82.3	2,137.3	77.3
Group 2	m	82.3	2,137.3	77.3
Group 3	m	137.2	2,137.3	77.3
Dry weight	kg/m	453.3	86.5	453.3
Wet weight	kg/m	394.4	68.5	394.4
Mbl uncorroded	kN	18,886	15,790	18,886
Mbl corroded	kN	16,373	NA	NA
Added mass coef.		1.0	1.0	1.0
Drag coef. (cd)		2.45	1.2	2.45

Table 9-2. As-Built Mooring Line Pretension

Line Number	Unit	Pretension
1	kN	2,348.5
2	kN	2,348.5
3	kN	2,357.4
4	kN	2,682.1
5	kN	2,699.9
6	kN	2,708.8
7	kN	2,833.4
8	kN	2,846.7
9	kN	2,846.7

Table 9-3. Riser Properties

Line Properties	Unit	SCR (Gas Export)	SCR (Oil Export)	TTR (Prod.)
Diameter	mm	273.0	324.0	324.0
Dry weight	kg/m	112.0	176.1	181.1
Wet weight	kg/m	51.9	91.7	96.7
Pretension	kN	1,290	2,224	3,585
Added mass coef.		1.0	1.0	1.0
Drag coef. (cd)		1.0	1.0	1.0

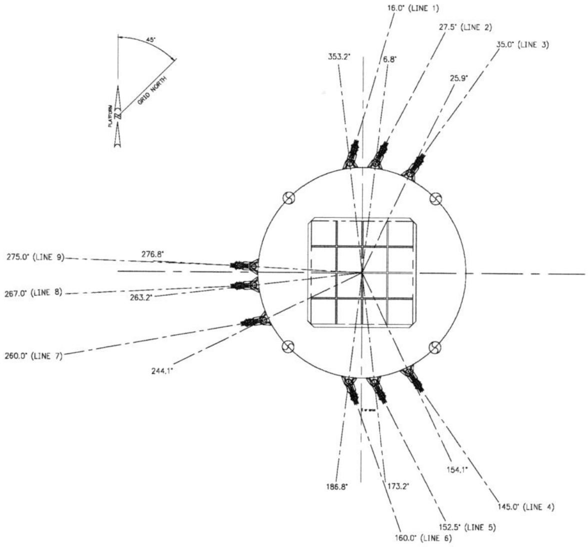

Fig. 9-2. Mooring Configuration

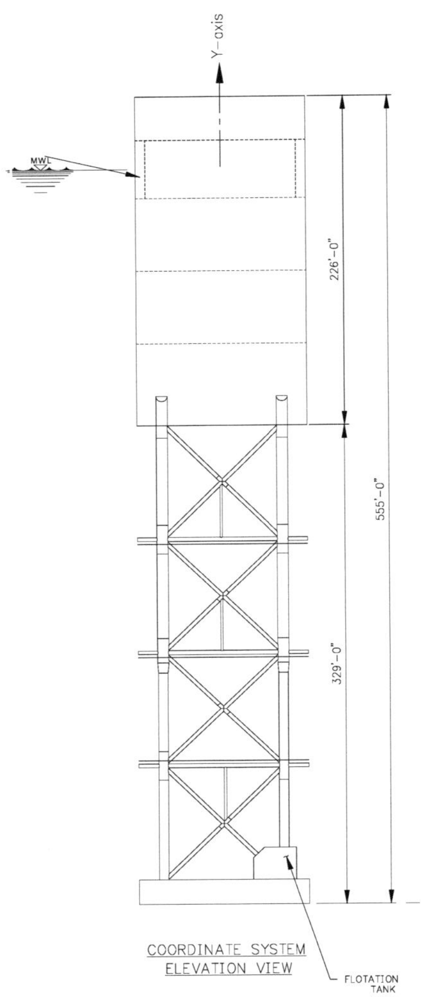

Fig. 9-3. Hull Configuration

The mass properties of the spar depend on the number of TTRs and SCRs installed and on the variable load of the topsides. Tables 9-4 and 9-5 show estimates of mass and hydrostatic properties for the two events being analyzed.

Table 9-4. Mass Properties

Mass Properties	Unit	Isidore (09/25/02)	Summer storm (08/15/03)
Weight	kN	507,010	492,665
KG	m	102.1	105.8
Vertical mooring load	kN	21,217	21,217
Vertical scr load	kN	3,340	3,340
TTR load at keel	kN	21,511	35,851
Roll gyradius	m	61.0	61.0
Pitch gyradius	m	61.0	61.0
Yaw gyradius	m	12.5	12.5

Table 9-5. Hydrostatic Properties

Hydrostatic Properties	Unit	Value
Draft	m	153.9
Displacement	kN	553,073
KB	m	107.7
Water plane area	m2	819.4

The weight and KG include flooded water entrained in the soft tank and the center well.

Numerical Modeling of Spar Hull, Mooring, and Risers

MLTSIM Modeling

MLTSIM provides various options for managing hydrodynamic loads, for example, combined Morison members (for the truss) and a panelized method (for the hard tank). The choice is normally based on the level of detail required and the diameter of the spar relative to the wavelength to be analyzed.

Figure 9-4 shows a schematic of an MLTSIM model of the Horn Mountain spar using Morison members. The hard tank, truss, and soft tank are modeled as slender bodies. The heave plate is represented as multiple slender members with appropriate hydrodynamic coefficients to capture the drag and added mass effects. The analysis employed Morison drag elements to model viscous forces on slender bodies. This model is consistent with the one used during the design phase for survival conditions. The Morison drag coefficients are given in Table 9-6.

Table 9-6. Drag Coefficients

Member	Drag Coef.
Hard tank lateral	1.2
Hard tank axial	1.0
Truss legs	1.1
Truss bracing	1.1
Heave plate	11.6
Soft tank lateral	0.89
Soft tank vertical	3.0

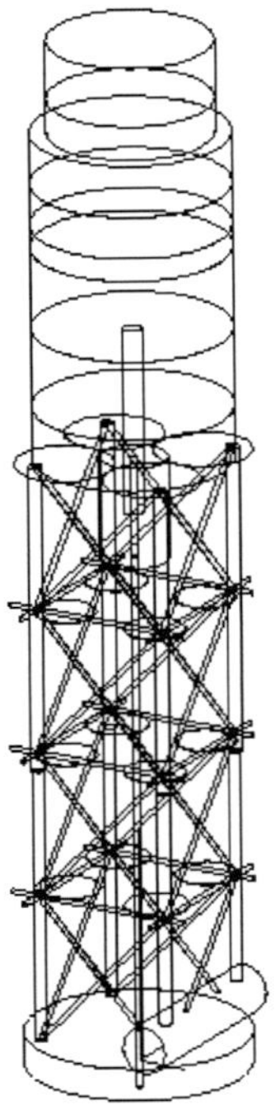

Fig. 9-4. Morison members in MLTSIM

WINPOST Modeling

The WINPOST model is based on a hybrid model of Morison members and a panelized body. Morrison elements are used to model the hydrodynamic force of slender bodies, such as the truss. The force on the hard tank, heave plates, and soft tank are calculated by the first-order diffraction/radiation program WAMIT (Lee et al. 1999). Viscous forces are obtained from Morison drag elements.

WAMIT is also used to calculate the added mass and hydrodynamic damping, first-order wave-frequency forces, and second-order mean. The second-order diffraction/radiation effects are obtained using Newman's approximation (Newman 1974). that is, the off-diagonal components of the second-order difference-frequency quadratic transfer functions (QTFs) are approximated by their diagonal values (mean drift forces and moments). The approximation can be justified only when the relevant natural frequency and the slope of QTFs near the diagonal are very small.

The sum-frequency parts are not important for spar motion analysis and are not included in the subsequent motion analysis. All the hydrodynamic coefficients were calculated in the frequency domain and the corresponding forces converted to the time domain using a two-term Volterra series expansion (Ran and Kim 1997). Frequency-dependent radiation damping was included in the form of a convolution integral in the time-domain simulation.

WINPOST is also used to calculate the hull/mooring/riser coupled statics/dynamics. An extension of the theory developed for slender rods by Garrett (1982) and Nordgren (1974) allows static/dynamic analysis of mooring lines and risers. Assuming that there is no torque or twisting moment, a linear momentum conservation equation can be derived with respect to a position vector $\vec{r}(s,t)$, which is a function of arc length s and time t:

$$-(B\vec{r}'')''+(\lambda\vec{r}')'+\vec{q}=m\ddot{\vec{r}} \tag{9-1}$$

$$\lambda = T - B\kappa^2 \tag{9-2}$$

where primes and dots denote the s-derivative and time-derivative, respectively, B is the bending stiffness, T the local effective tension, κ the local curvature, m the mass per unit length, and $\vec{q}$ the distributed force on the rod per unit length. The scalar variable λ can be regarded as a Lagrange multiplier. The rod is assumed to be elastic and extensible, thus the following condition is applied

$$\frac{1}{2}(\vec{r}'\cdot\vec{r}'-1)=\frac{T}{A_t E}\approx\frac{\lambda}{A_t E} \tag{9-3}$$

where E=Young's modulus, $A_t = A_e - A_i$(= outer – inner cross sectional area). Geometric non-linearity is fully considered for these equations, and there is no special assumption made concerning the shape or orientation of lines. The benefit of this equation is that it is directly defined in the global coordinate system and does not require any transformations to the local coordinate system.

The normal component of the distributed external force on the rod per unit length, q_n, is calculated using a generalized Morison equation:

$$q_n = C_I \rho A_e \dot{v}_n + C_D \frac{1}{2}\rho D|v_{nr}|v_{nr} + C_m \rho A_e \ddot{r}_n \tag{9-4}$$

where C_I, C_D, and C_m are inertia, drag, and added mass coefficients, and $\dot{v}_n, v_{nr}$, and $\ddot{r}_n$ are normal fluid acceleration, normal relative velocity, and normal structure acceleration, respectively. The symbols ρ and D are fluid density and local diameter. The effective weight, or net buoyancy, of the rod should be included in q_n as a static load.

A finite element method similar to Garrett (1982) was developed to solve the above mooring dynamics problem. Details of the methodology are given in Ran and Kim (1997). The finite element method (FEM) allows any combination of mooring types and materials as long as their deformations are small and within proportional limits. The upper ends of the mooring lines and risers are connected to the hull fairlead through generalized elastic springs and dampers. The combination of linear and torsional springs can model arbitrary connection conditions. The forces and moments proportional to the relative displacements are transmitted to the hull at the connection points. The transmitted forces from mooring lines and risers to the platform are given by

$$\tilde{F}_P = \tilde{K}(\tilde{T}\tilde{u}_P - \tilde{u}_I) + \tilde{C}(\tilde{T}\dot{\tilde{u}}_P - \dot{\tilde{u}}_I) \tag{9-5}$$

where $\tilde{K}, \tilde{C}$ are stiffness and damping matrices of mooring lines at the connection point, and $\tilde{T}$ represents a transformation matrix between the platform origin and connection point. The symbols $\tilde{u}_P, \tilde{u}_I$ represent column matrices for the displacements of the platform and connection points.

The interaction between the mooring line/riser and seafloor is modeled as an elastic foundation in the vertical direction. A continuous quadratic-spring stiffness is assumed, while the tangential friction between mooring line/risers is modeled as linear drag coefficient. In a three-dimensional problem with the origin of the coordinate system located on the mean water level and z-axis pointing upward, the interaction force on the line from seafloor can be expressed as

$$f_1 = 0,\ f_2 = 0,\ f_3 = c(r_3 - L)^2 \text{ for } r_3 - L < 0 \tag{9-6}$$

$$f_3 = 0 \text{ for } r_3 - L \geq 0$$

where L is the water depth, r_3 is the z-component of the line position vector r, and c is spring stiffness.

The following hull response equation can be combined into the riser/mooring-line equations in the time domain:

$$(\widetilde{M} + \widetilde{M}_a(\infty))\ddot{\widetilde{u}}_p + \int_0^\infty \widetilde{R}(t-\tau)\dot{\widetilde{u}}_p d\tau + \widetilde{K}_H \widetilde{u}_p = \widetilde{F}_D + \widetilde{F}^{(1)} + \widetilde{F}^{(2)} + \widetilde{F}_p + \widetilde{F}_w + \widetilde{F}_c + \widetilde{F}_{WD} \tag{9-7}$$

where $\widetilde{M}, \widetilde{M}_a$ are mass and added mass matrix, $\widetilde{R}$=retardation function (inverse cosine Fourier transform of radiation damping) matrix, $\widetilde{K}_H$= hydrostatic restoring coefficient, $\widetilde{F}_D$= drag force matrix on the hull, $\widetilde{F}^{(1)}$, $\widetilde{F}^{(2)}$ = first- and second-order wave load matrix on the hull, $\widetilde{F}_p$= transmitted force matrix from the interface, $\widetilde{F}_w$= dynamic wind loading, $\widetilde{F}_c$= current loading on hull, and $\widetilde{F}_{WD}$ = wave drift damping force matrix.

The added mass at infinite frequency was obtained from the Kramers-Kronig relation. A two-term Volterra series was used (Kim and Yue 1990) for the time series of $\widetilde{F}^{(1)}, \widetilde{F}^{(2)}$, and $\widetilde{F}_{WD}$. The hull drag force in the normal direction was calculated for the instantaneous hull position based on the Morison drag formula with relative velocity squared.

Newton's iterative method provided a solution for the static problem of the integrated system. The dynamic problem was integrated using an efficient and reliable time-marching scheme similar to the Adams-Moulton method (Garrett 1982).

In the dynamic program, special consideration is required because the time derivatives of λ do not appear in the equations, and the added mass matrix is a function of the instantaneous position. The free-surface fluctuation and possible contact of mooring lines and catenary risers with the seafloor require special consideration.

In the frequency domain analysis, a mean position is first estimated based on the given static loads—wind, current, and wave drift forces. Next, the dynamic analysis with linearized drag forces is performed with respect to the mean position. The spectral analysis for the first- and second-order wave forces is explained in detail in Kim and Yue (1991).

To determine the stochastic linearization of the nonlinear drag force in the presence of currents, the first calculation is

$$C_d(v_{nr}+c_n)\,|\,v_{nr}+c_n\,|=C_L v_{nr}+F_{mean}, \tag{9-8}$$

Where c_n is the normal current velocity. After minimizing the difference between the right-hand and left-hand side with respect to C_L and F_{mean}, the following can be obtained:

$$C_L=\frac{E[C_d v_{nr}(v_{nr}+c_n)\,|\,v_{nr}+c_n\,|]}{E[v_{nr}v_{nr}]} \tag{9-9}$$

and

$$F_{mean}=E[C_d(v_{nr}+c_n)\,|\,v_{nr}+c_n\,|] \tag{9-10}$$

The two coefficients can be determined by assuming Gaussian random process, as explained in Rodenbusch et al. (1986).

The spar hull, heave plate, and soft tank are discretized by 6444 panels as shown in Figure 9-5. Each mooring line, SCR, and TTR are modeled by 13, 15, and 10 high-order elements, respectively. The convergence of the discretization of the mooring lines and risers was checked by doubling the number of elements for selected lines. The mooring lines and SCRs are assumed to be hinged at the fairlead and sea floor. The TTRs are hinged at the sea floor but held at the spar keel by a constant force through buoyancy cans. Therefore, the TTR tension is not included in the vertical static equilibrium of the hull.

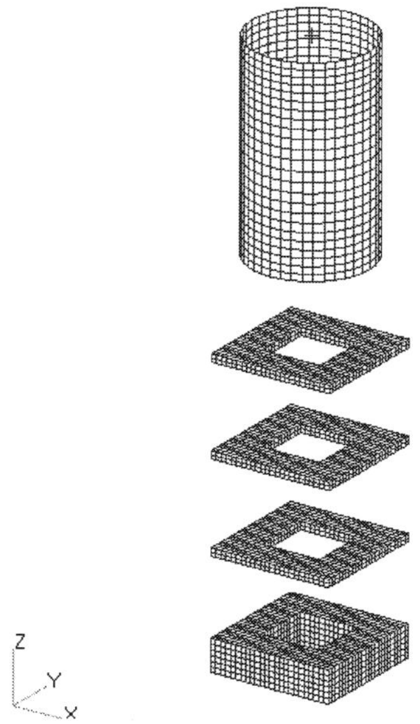

Fig. 9-5. Descretization of hull, heave plate, and soft tank

Metocean

This study evaluates Hurricane Isidore, which swept through the Gulf of Mexico in September 2002, and a summer storm that occurred on August 15 2003. Metocean data for the storms are shown in Table 9-7. Figures 9-6 and 9-7 show their directionalities.

The current is assumed to be steady and the irregular waves unidirectional. A JONSWAP spectrum with significant wave height (Hs) and peak period (Tp) tabulated in Table 9-7 is calibrated against the field measurements. Figures 9-8 and 9-9 show the comparison between the computed and best-fit JONSWAP spectrum. The API wind spectrum is used to generate time-varying wind forces.

Table 9-7. Metocean Condition

Description	Isidore		August Storm	
Wave (JONSWAP)				
Hs (m)	6.36		2.84	
Tp (sec)	12.2		8.2	
Shape	1.3		2.4	
Wind (API)				
V (1hr@10 m) - m/s	19.38		11.55	
Peak Coefficient (fp)	0.025		0.025	
Current	Depth (m)	Vel (m/s)	Depth (m)	Vel (m/s)
	0.00	0.26	0.00	0.15
	43.59	0.20	60.96	0.21
	59.74	0.13	76.20	0.26
	75.59	0.25	137.16	0.31
	107.59	0.19	182.88	0.27
	171.60	0.10	243.84	0.20
	251.76	0.04	281.94	0.18
	363.63	0.07	350.52	0.20
	1645.9	0.04	1645.9	0.10

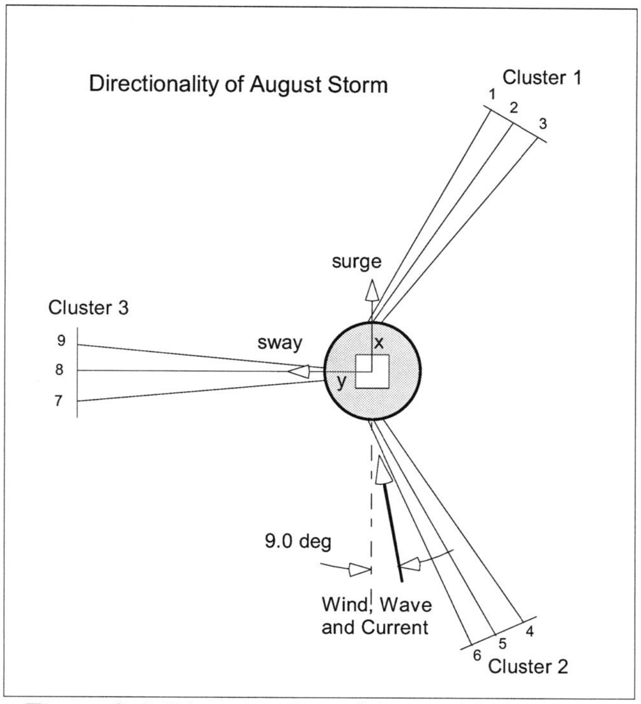

Figure 9-6. Directionality of August 2003 storm

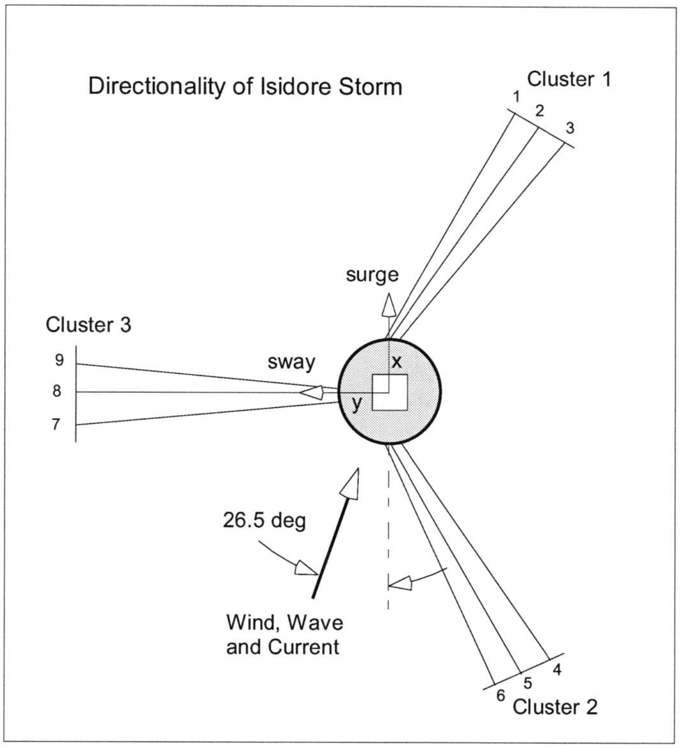

Fig. 9-7. Directionality of Hurricane Isidore

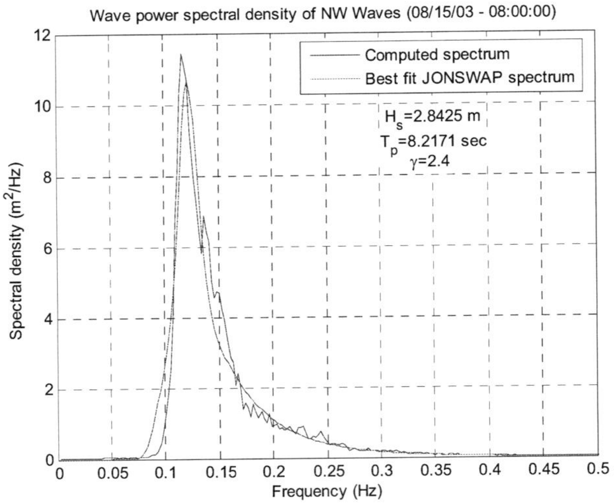

Fig. 9-8. Wave power spectral density during August storm

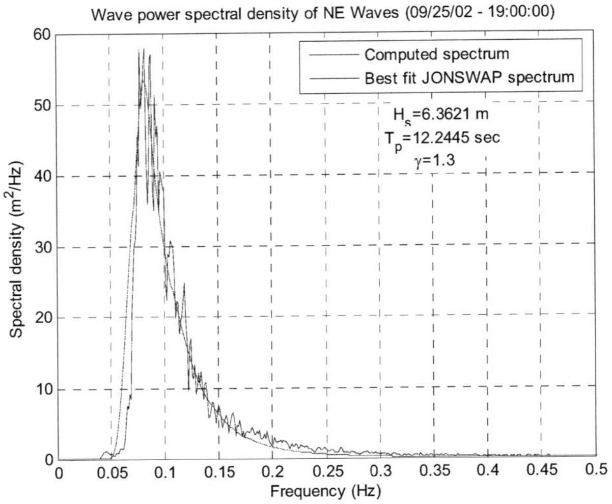

Fig. 9-9. Wave power spectra density during Hurricane Isidore

Results and Analysis

Static-Offset Simulations

The surge and sway static offset simulation was conducted by applying a force at the VCG (vertical center of gravity) in the horizontal direction in calm water. Figures 9-10 and 9-11 show the static offset results in surge and sway, respectively.

The surge static-offset results show a small discrepancy between MLTSIM and WINPOST, while results in sway match very well. This discrepancy is believed to be caused by the different approaches to modeling TTRs. MLTSIM accounts for the effect of TTRs by simulating a massless element with a soft spring, while WINPOST simulates this effect by taking into account the distributed weight of the TTR. When the spar is laterally displaced, the TTR departure angle in WINPOST is higher than the TTR departure angle in MLTSIM, which makes the restoring force smaller.

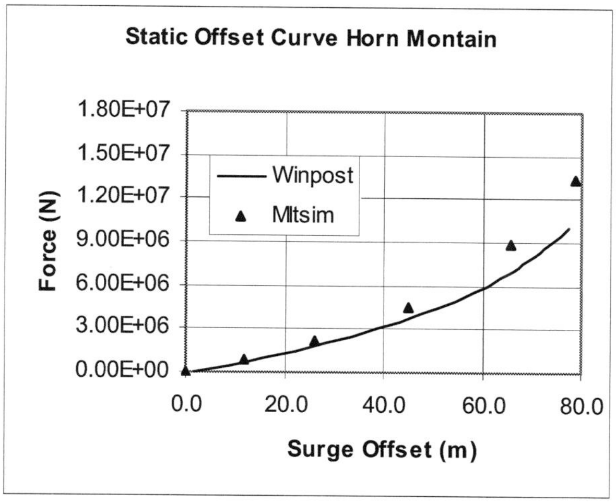

Fig. 9-10. Static offset curve in surge

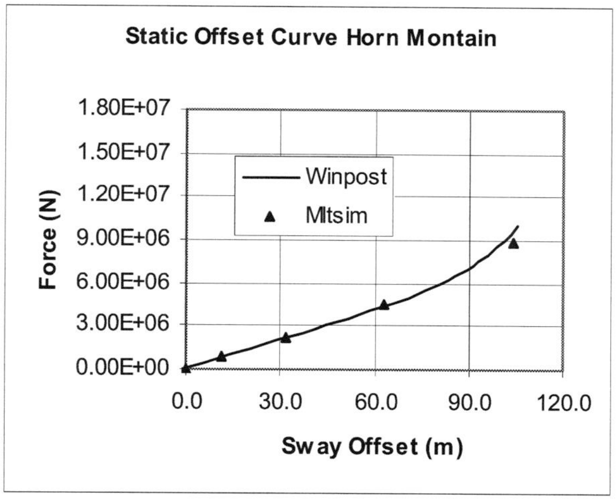

Fig. 9-11. Static offset curve in sway

Free-Decay Simulations

Natural periods and damping in calm water are determined from a simulation of the decay responses when initial conditions are not in equilibrium. The results in surge, heave, and pitch are shown in Figures 9-12, 9-13, and 9-14. The natural heave period derived using MLTSIM is slightly higher than that derived using WINPOST. The possible cause of this deviation is the way the heave plate added mass is modeled.

In MLTSIM, heave plate added mass is calibrated against model tests (Prislin et al. 1998). which takes into account a portion of added mass from the area containing riser guides. WAMIT does not model this area in order to avoid undesirable numerical problems. The added mass computed using WAMIT is slightly lower than that of the model tests, which result in a lower natural period in WINPOST.

The pitch-decay test shows that the natural period results from MLTSIM and WINPOST match very well, while damping is appreciably different. The difference is due to additional damping from mooring lines and risers, which is not modeled by MLTSIM. These results indicate that the damping could be important for surge/sway and pitch/roll responses in deep water.

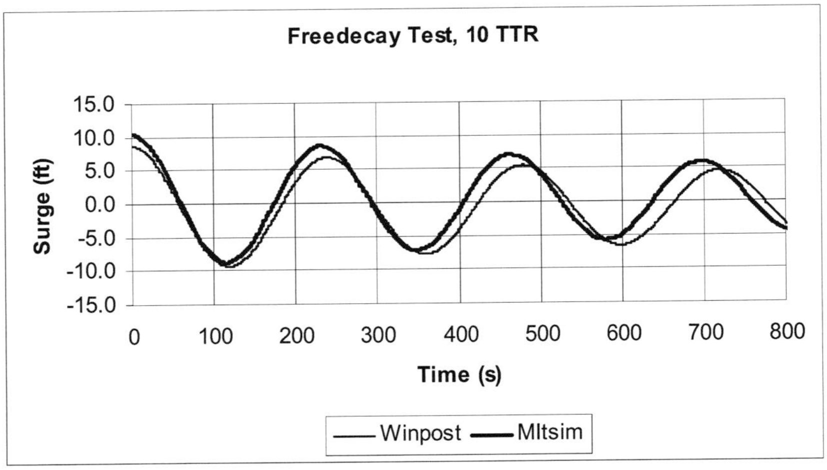

Fig. 9-12. Surge free-decay test

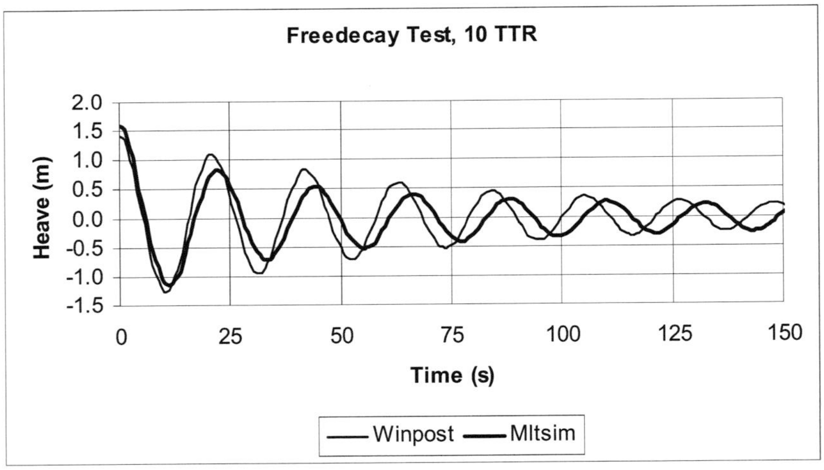

Fig. 9-13. Heave free-decay test

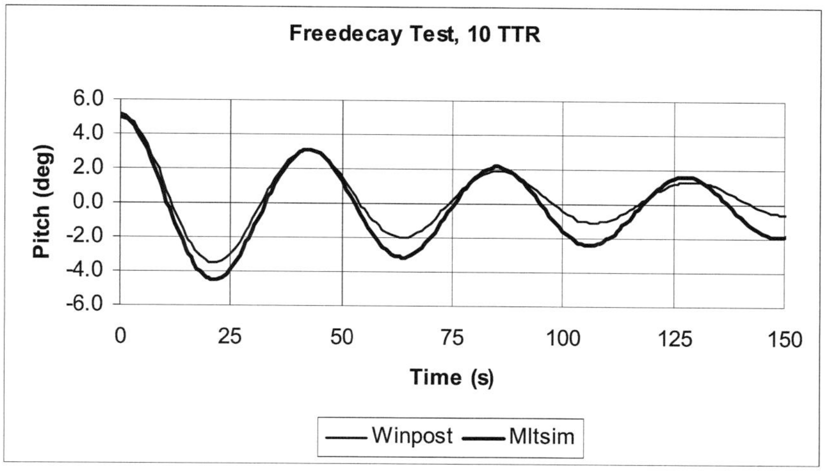

Fig. 9-14. Pitch free-decay test

Two Numerical Methods in Time Domain, Quasi-Statically Coupled and Dynamically Coupled, with Field Measurement During August Storm

Quasi-static coupled analyses were performed using MLTSIM, while dynamic coupled analyses were carried out with WINPOST. Figures 9-15 to 9-19 summarize the statistical results of the August storm analysis. All results represent a single 3-hour realization of the environmental conditions. Surge motions shown here are calculated at the cellar deck, approximately 21 m above the waterline. MLTSIM slightly under-predicted the mean offset, presumably because of the stiffer mooring characteristics previously discussed (Figure 9-10).

The dynamic values for surge, Figure 9-15, show computed values that are consistently higher than measured values, especially for the low-frequency components. This is also the case for pitch motions (Fig. 9-17 and 9-18). Note that surge at the cellar deck includes a pitch component, so these results are not independent.

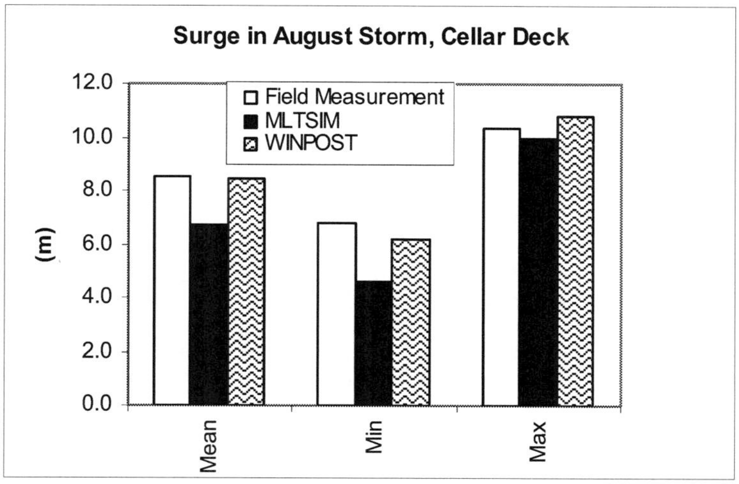

Fig. 9-15. Surge extreme, August storm

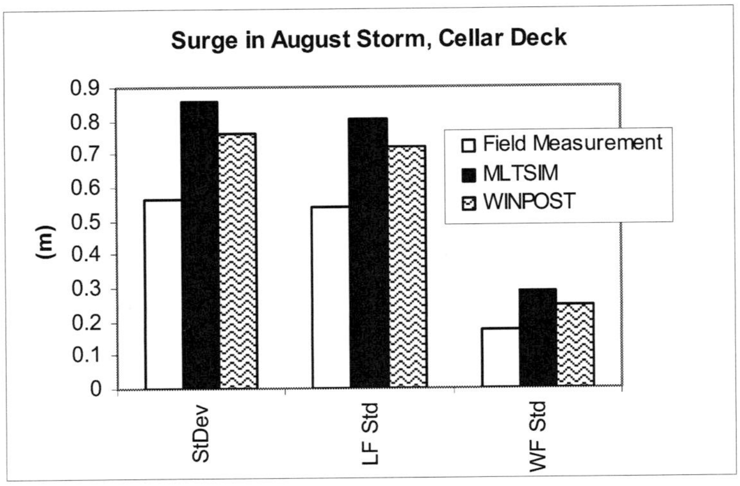

Fig. 9-16, Surge dynamics, August storm

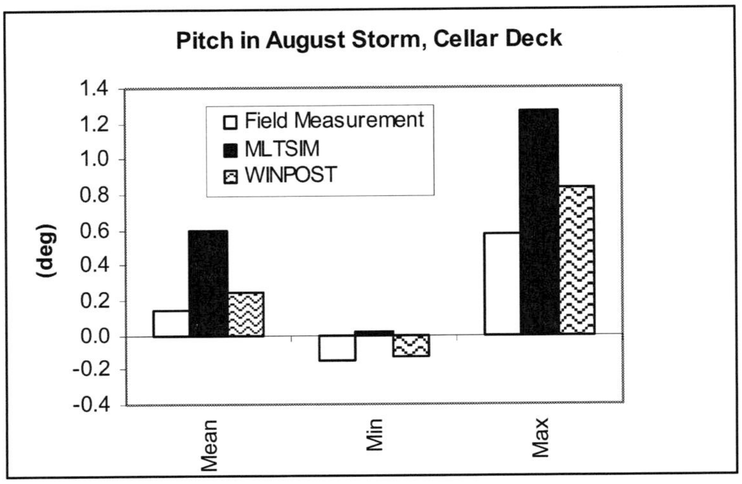

Fig. 9-17. Pitch extreme, August storm

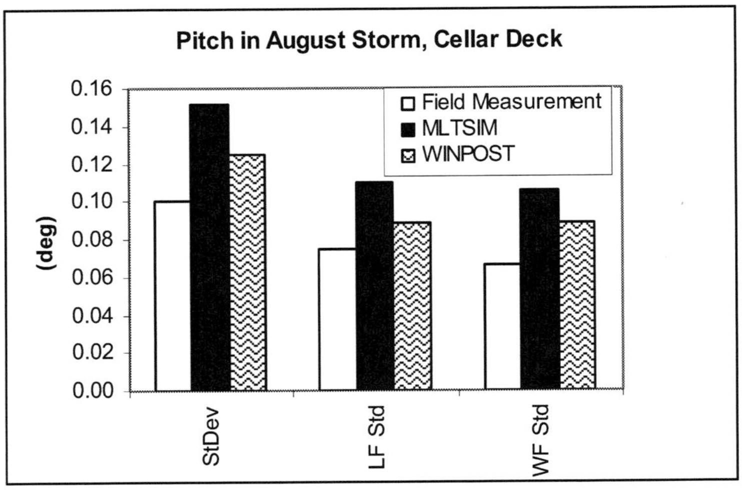

Fig. 9-18. Pitch dynamic, August storm

The higher low-frequency surge and pitch motions computed by MLTSIM are attributed to lower damping. The higher wave-frequency response cannot be caused by damping, but it could be due to the Morison model used by MLTSIM versus the radiation/diffraction model used by WINPOST. This hypothesis is supported by the analysis results for Hurricane Isidore, where the wave-frequency responses generated using the programs are more consistent. These results are expected for a storm with wave energy concentrated at a longer wavelength where the slender body theory and diffraction/radiation theory should converge.

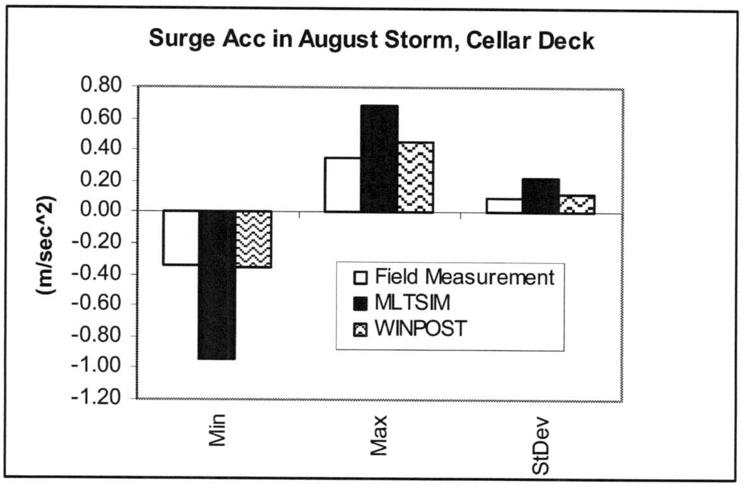

Fig. 9-19. Cellar deck acceleration, August storm

Figure 9-19 illustrates the comparison of surge accelerations. The results include the gravitational component parallel to the deck [g*sin(pitch)], so there is a low-frequency component factored into these results. The WINPOST coupled model with diffraction/radiation effects clearly yields results that are closer to the actual measurements.

Interestingly, both methods seem to over-predict surge and pitch results. One possible reason for this is the assumption used in the analysis that the waves are unidirectional and long crested. In reality, wind, waves, and current were coming from different directions.

Two Numerical Methods in Time Domain, Quasi-Statically Coupled, and Dynamically Coupled, with Field Measurement During Hurricane Isidore

Figures 9-20 to 9-24 present the statistical summaries for simulations of Hurricane Isidore. Here the surge results are more consistent, though MLTSIM still appears to over-predict low-frequency pitch. Wave frequency pitch is the same for both methods.

Figure 9-24 shows how the lateral accelerations again reflect the tendency of the uncoupled analysis (MLTSIM) to over-predict pitch responses, which in turn results in over-prediction of the accelerations.

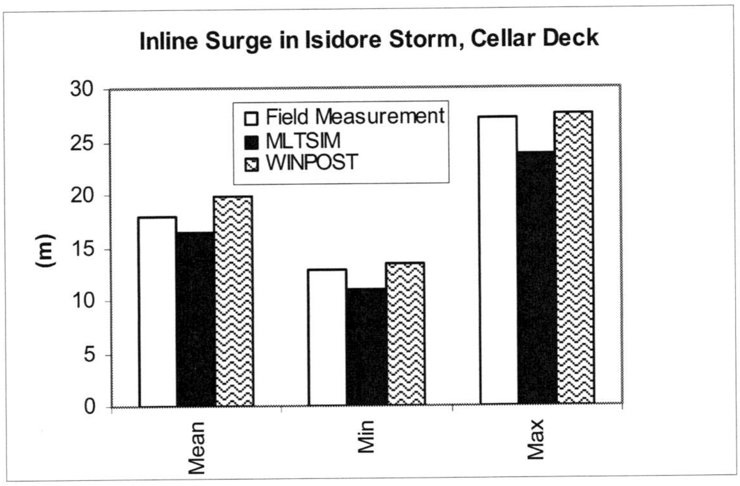

Fig. 9-20. In-line surge extreme, Hurricane Isidore

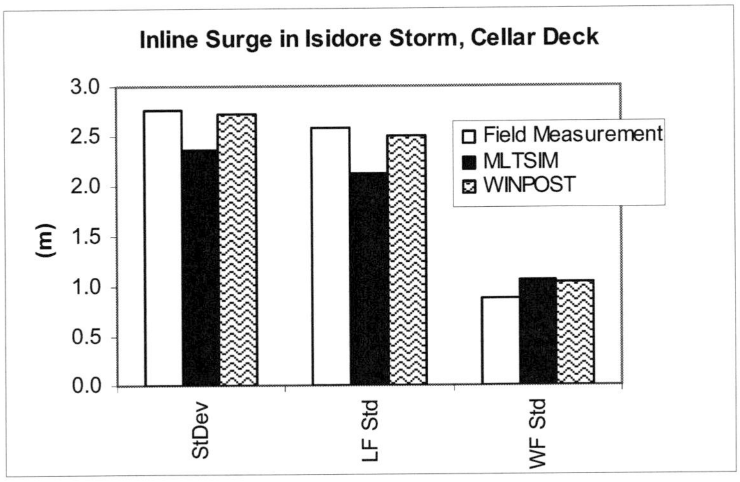

Fig. 9-2.1 In-line surge dynamics, Hurricane Isidore

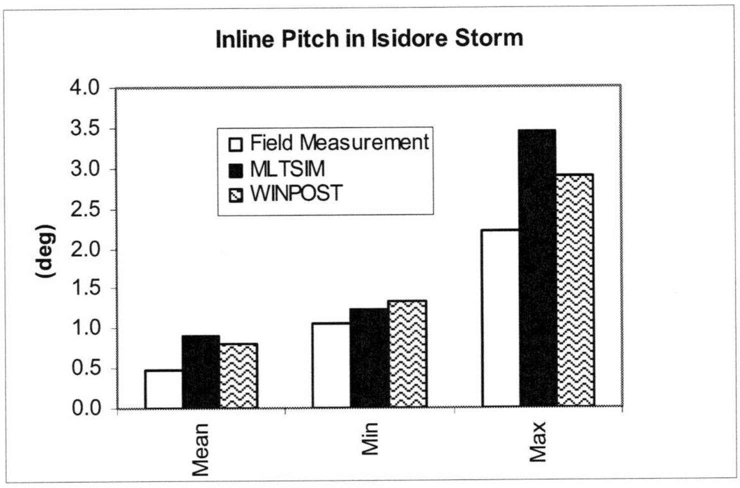

Fig. 9-22. In-line pitch extreme, Hurricane Isidore

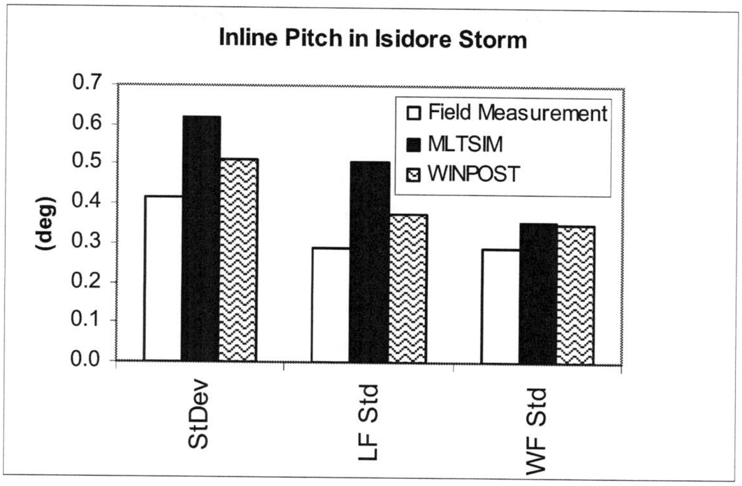

Fig. 9-23. In-line pitch dynamics, Hurricane Isidore

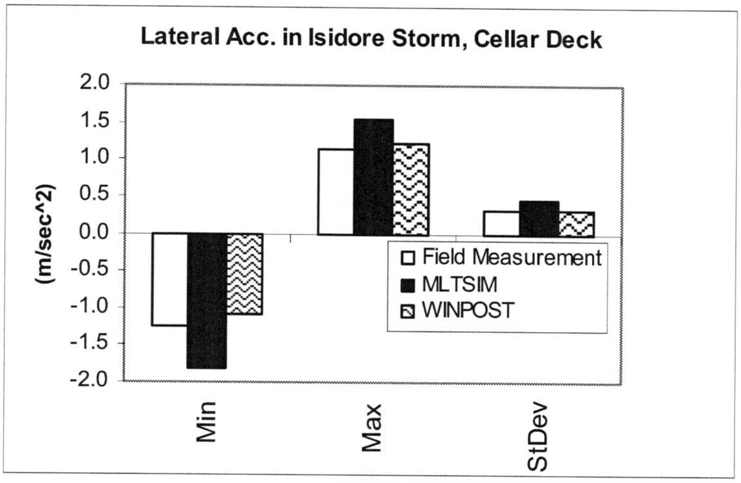

Fig. 9-24. Cellar deck acceleration, Hurricane Isidore

Figures 9-25 to 9-27 show the spectral densities of the responses for surge, pitch, and yaw. The results clearly show the effect of coupled analysis in terms of damping, particularly in pitch and yaw.

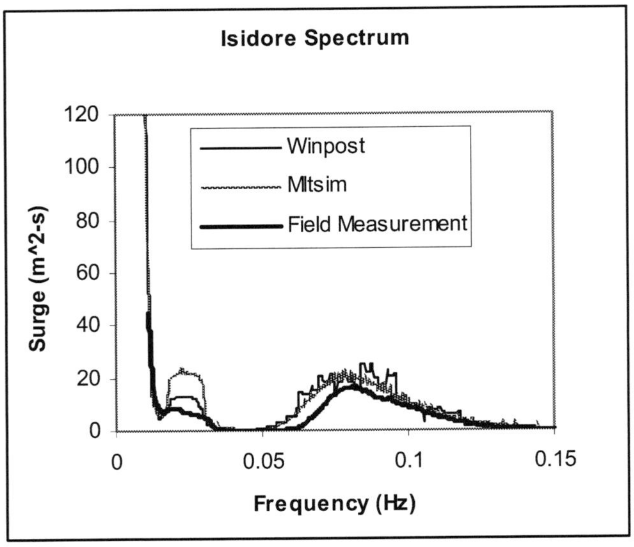

Fig. 9-25. Surge spectrum density, Hurricane Isidore

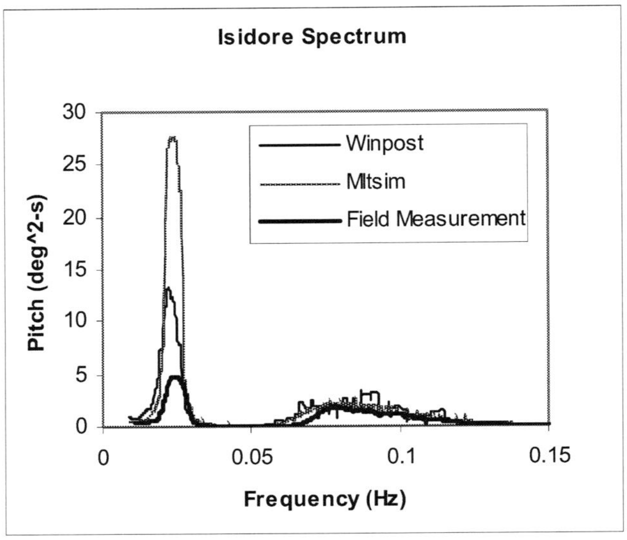

Fig. 9-26. Pitch spectrum density, Hurricane Isidore

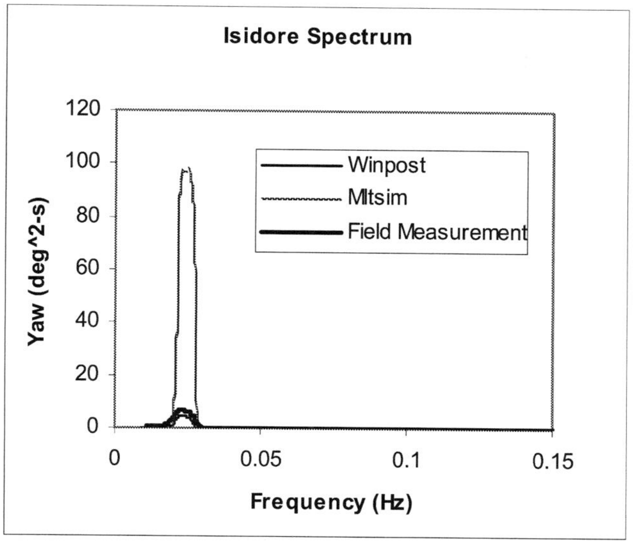

Fig. 9-27. Yaw spectrum density, Hurricane Isidore

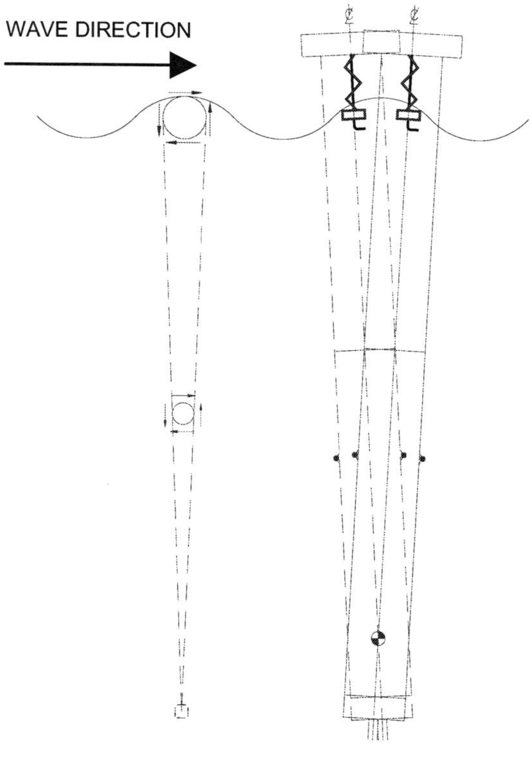

Fig. 9-28. Spar wave-frequency pitch response

The spar's response at the riser hangoff location is particularly interesting. TTRs and SCRs are typically attached near the bottom of the spar or at the keel. Figure 9-28 illustrates the wave-frequency response in pitch, where the center of rotation tends to be close to the keel. The placement of the center of rotation results in very small surge motions at the keel.

The measured responses of the Horn Mountain spar during Hurricane Isidore were transferred to the keel to verify this effect. The results appear in Figures 9-29 and 9-30. Of particular interest is the comparison between the standard deviation of the wave-frequency in-line response at the cellar deck (Figure 9-21) and that of the response at the keel (Figure 9-30): 3 ft. vs. 0.3 ft. (0.9 m vs. 0.09 m).

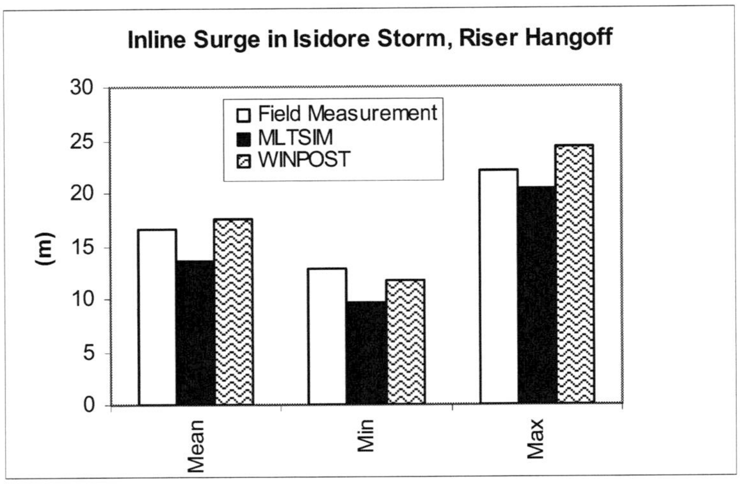

Fig. 9-29. In-line surge extreme at riser hangoff, Hurricane Isidore

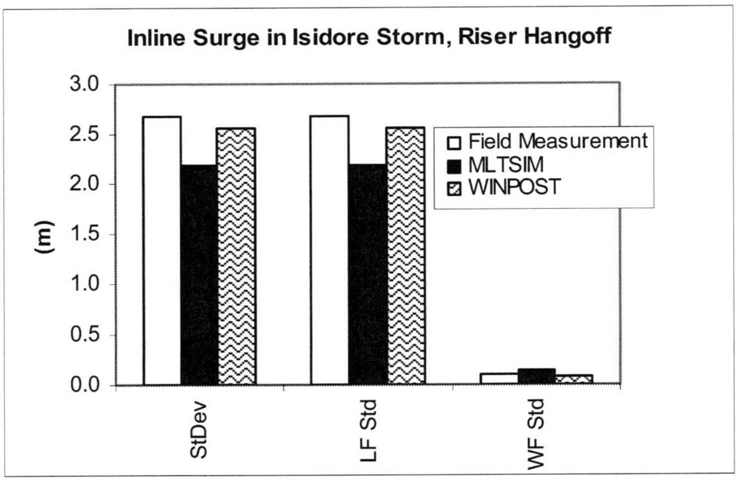

Fig. 9-30. In-line surge dynamics at riser hangoff, Hurricane Isidore

Technical challenges and economics dictated that sensing devices to measure mooring tension be installed right after the chain jack or chain stopper so that measured data represent tensions at this location. Numerical simulations, on the other hand, obtained mooring tensions at the fairlead, which is about 66.75 m below the chain jack. To make a fair comparison, tensions from simulations need to be corrected to incorporate static and dynamic frictions from the fairlead bearing as well as the gravity weight of the mooring chain above the fairlead.

Figure 9-31 shows a side view of a fairlead. The mooring chain is wrapped securely around the roller, which rotates to the bearing. Dynamic friction forces begin to act when the difference between inboard and outboard tensions is greater than the static friction forces. Equations 9-11 and 9-12 show the formulations used to compute dynamic frictions.

$$\sum M_0 = 0$$
$$-\mu N r + T_{outboard} R - T_{inboard} R = 0 \qquad (9\text{-}11)$$

$$\sum F_x = 0$$
$$T_{outboard} \sin\gamma - N\cos(\tfrac{\gamma}{2}) - \mu N \sin(\tfrac{\gamma}{2}) = 0 \qquad (9\text{-}12)$$

where
μ = dynamic friction coefficients
R = guide roller radius
r = bearing radius
N = normal force at bearing contact
γ = departure angle of mooring line from vertical
$T_{inboard}$ = mooring tension inboard fairlead
$T_{outboard}$ = mooring tension outboard fairlead
M_0 = moment with respect to origin
F_x = force in x-direction

Combining eqs 9-11 and 9-12 results in the following relationship:

$$T_{inboard} = T_{outboard}\left[1 - \frac{\mu r}{R}\frac{\sin(\gamma)}{\cos\left(\frac{\gamma}{2}\right) + \mu\sin\left(\frac{\gamma}{2}\right)}\right] \qquad (9\text{-}13)$$

A simple post-processing program was developed to compute $T_{inboard}$ using $T_{outboard}$ and γ output time series from numerical simulations of the platform. The most difficult part of solving for $T_{inboard}$ was determining the static and dynamic friction coefficient of a bearing. Initial values of friction coefficients were based on the manufacturer's specifications for dynamic friction (0.1). Static friction data is not available. A sensitivity analysis comparing measured and calculated tension time series is used to establish the actual static and dynamic friction coefficients, which were found to be 0.06 and 0.05, respectively.

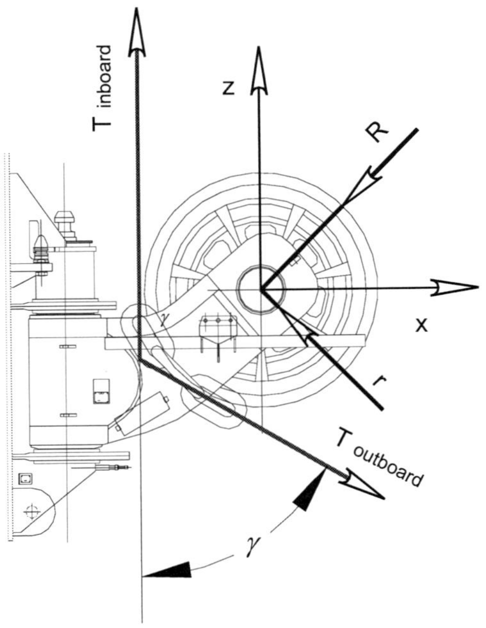

Fig. 9-31. Schematic drawing of a typical fairlead

In the following sections, tension at the chain jack is defined as the sum of $T_{inboard}$ and the weight of chain between the fairlead and the chain jack. Tension at the fairlead is the same as $T_{outboard}$.

Based on rigid body motions obtained from simulations, mooring tensions were compared to field measurements. Because the load-measuring devices to measure mooring tension are located at the chain jack, the results represent tension at the chain jack. A static correction of 60 kips was applied to the simulated tensions at the fairleads to account for the weight of the chains between the chain jacks and the fairleads.

Figures 9-32 to 9-34 present the mean, maximum, and standard deviation of mooring tensions at the chain jacks and fairleads. Fair comparisons have to be made between field measurement at the chain jack and results generated using WINPOST. Tension results computed using WINPOST and MLTSIM at the fairleads are also presented so that results using a fully coupled dynamic approach can be compared to those employing quasi-static analysis.

WINPOST standard deviation was over-predicted by a factor of two on most loaded lines, while the mean and maximum tension at the chain jack measured in the field matched the WINPOST results very well.

There are at least three factors that can explain the difference in the dynamic tensions:

- Uncertainty in the friction coefficients: higher friction coefficients reduce the tension standard deviation, which reduces the discrepancy.

  Directionality of the marine environment: simulations assume that wind, waves, and current come from a fixed direction. In reality, wind, waves, and current were coming from different directions. There was also a wind-direction shift every 10 minutes on average that changed wind direction about 10° to 30°. Shorter gusts showed more variability. The 20-second average gusts varied +/-20° from the mean direction. Figures 20-23 show results for in-line motions, which are the vector summation of surge and sway motions at a 26.5° heading from the x-axis (see Fig. 9-7).

- Mooring line pretension: tension standard deviation can be affected by the difference between the pretension used in the simulations and the actual pretensions.

The standard deviation of the tensions at the fairleads using the quasi-static approach (MLTSIM) is lower by a factor of up to four than results achieved using the coupled approach (WINPOST). This result is not unexpected because the quasi-static approach is well known to under-predict tension dynamics (Steen et al. 2003). In the design phase, quasi-static tensions are corrected for dynamic effects, which yield results similar to those generated using WINPOST.

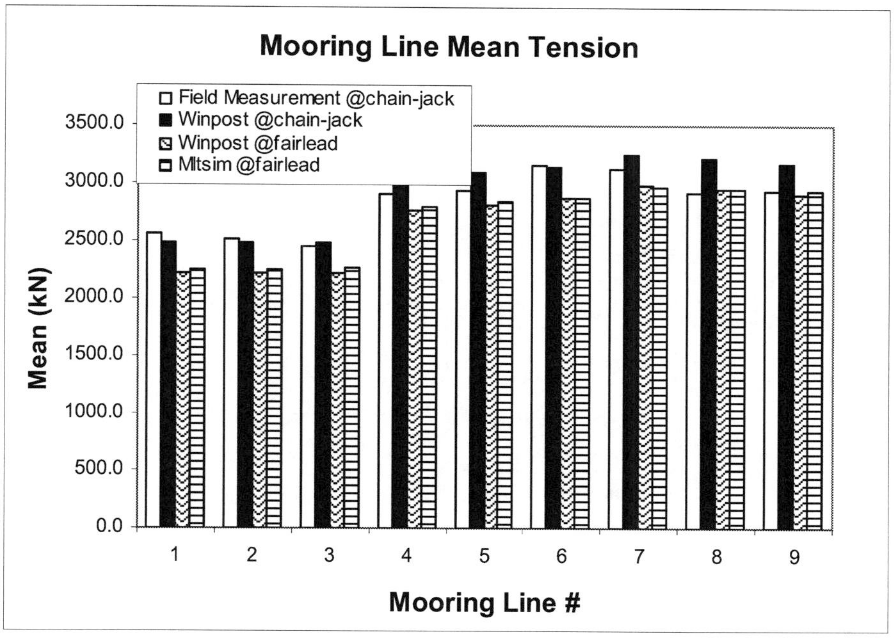

Fig. 9-32. Mean tension comparison, motion from simulation

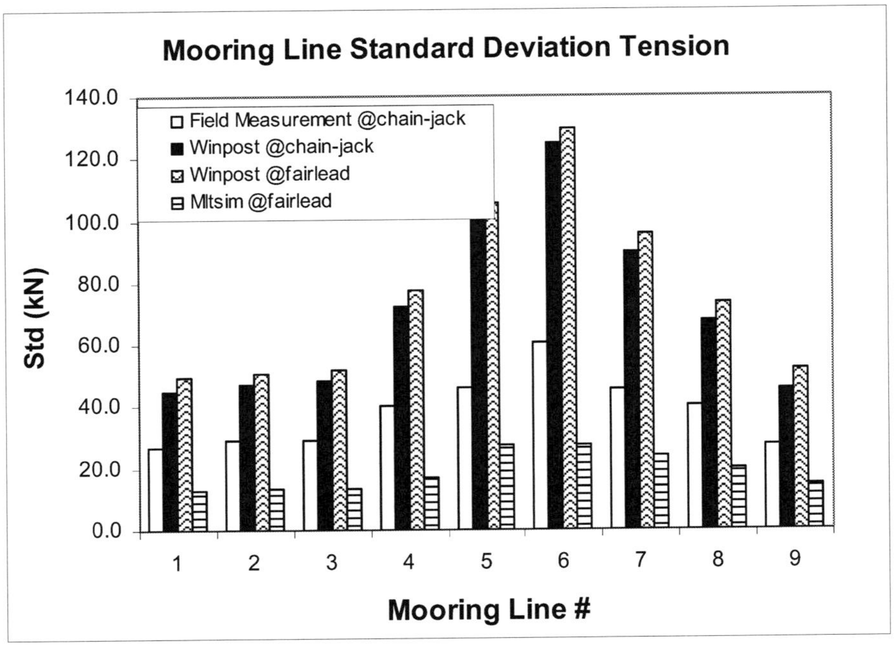

Fig. 9-33. Standard deviation tension comparison, motion from simulation

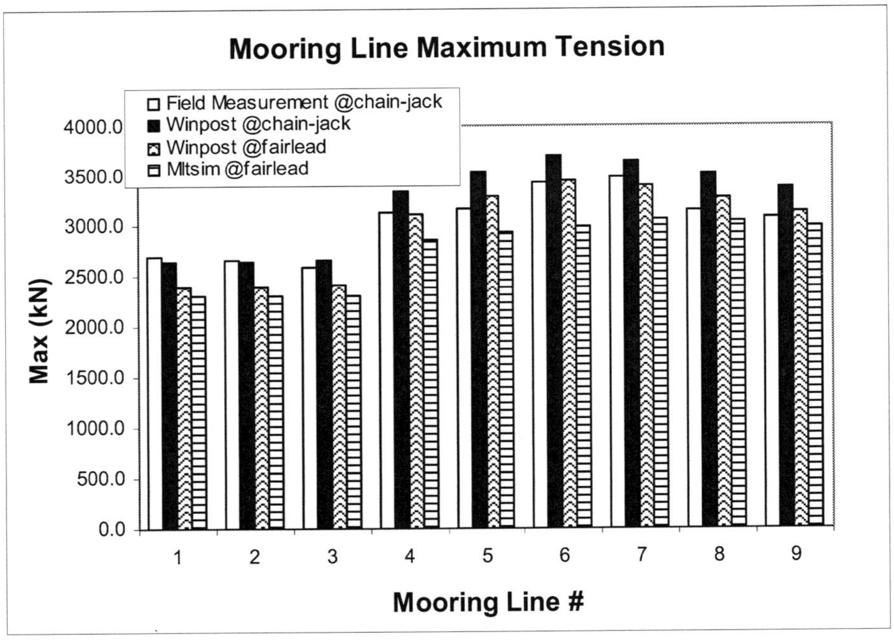

Fig. 9-34. Maximum tension comparison, motion from simulation

The following tension analysis computes mooring tensions using the measured platform motions as input to WINPOST. Figures 9-35 and 9-36 show that the maximum tensions matched very well, while the standard deviations were slightly different. The tension standard deviations at the chain jack for the mooring lines in cluster 1 (least loaded line) were under-predicted, but were over-predicted in cluster 3 (most loaded line). The possible discrepancy could have been the result of the initial simulated mooring pretensions differing from the actual pretensions.

Figures 9-37 and 9-38 show a further comparison between measured and simulated tensions, time history, and density spectrum for mooring line #7. The comparison shows that good agreement in the time and frequency domains is obtained with a static friction coefficient of 0.06 and a dynamic friction coefficient of 0.05.

Figure 9-39 compares maximum tension with and without friction. Ignoring fairlead friction results in a slightly conservative estimate for the tension at the chain jack especially for moorings in cluster 2. This procedure has been adopted as the standard practice in all spar designs.

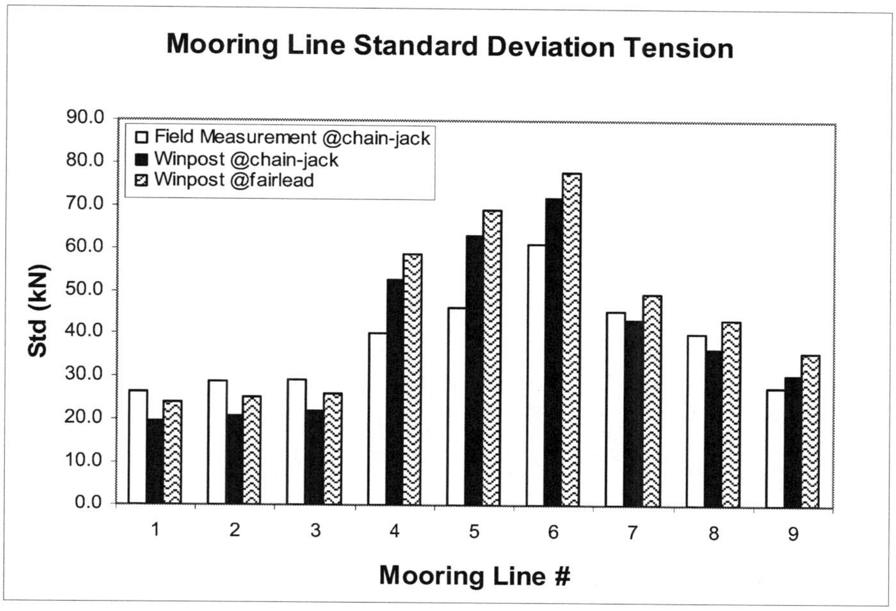

Fig. 9-35. Standard deviation tension comparison, motion from field measurement

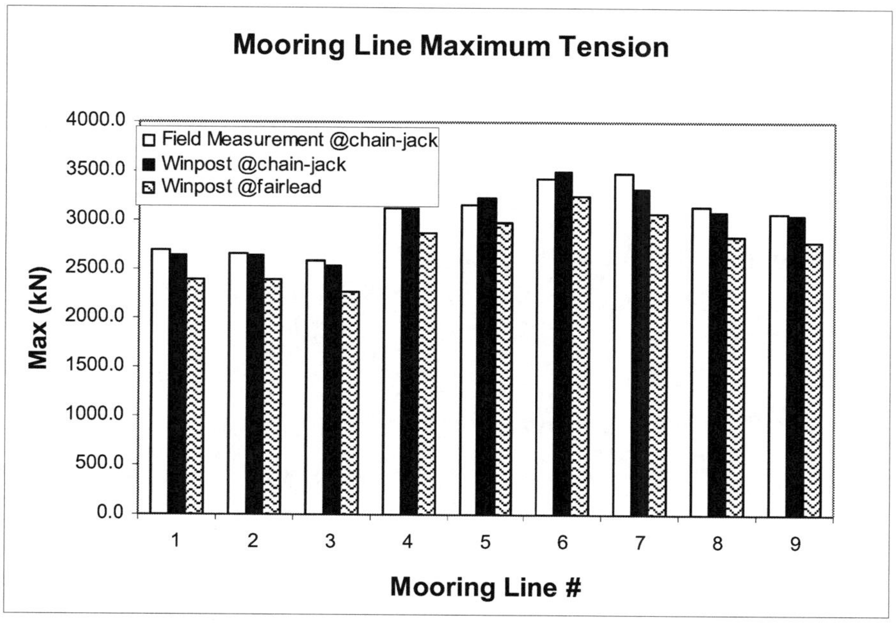

Fig. 9-36. Maximum tension comparison, motion from field measurement

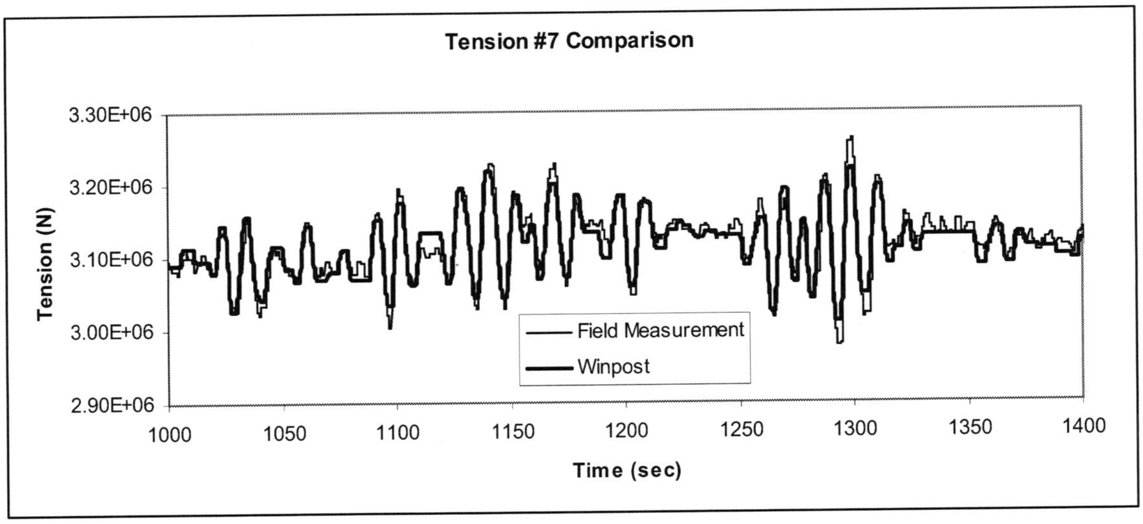

Fig. 9-37. Snapshot of tension comparison, motion from field measurement

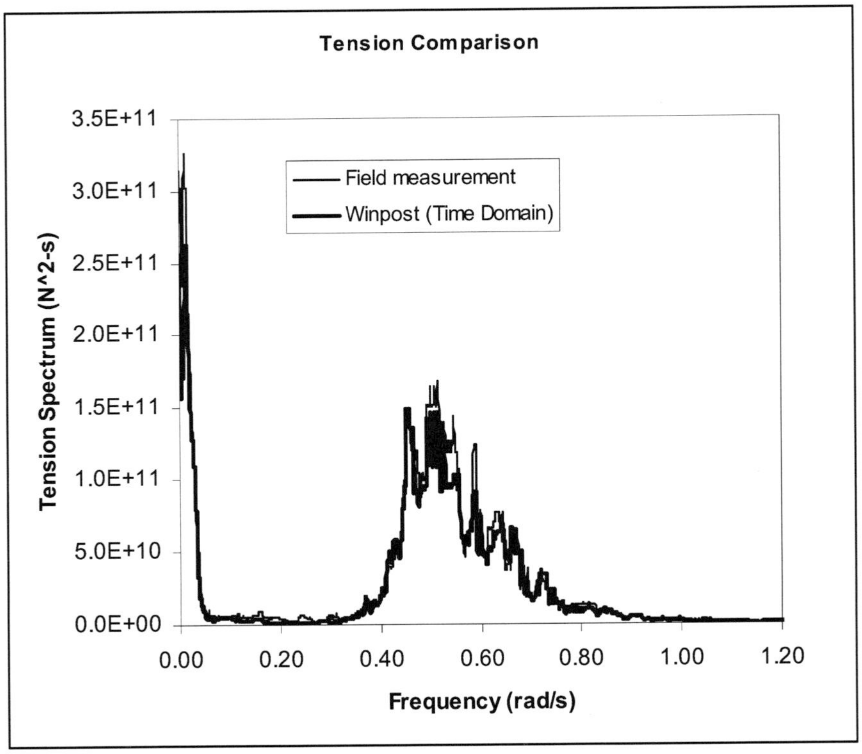

Fig. 9-38. Density spectrum of tension comparison, motion from field measurements

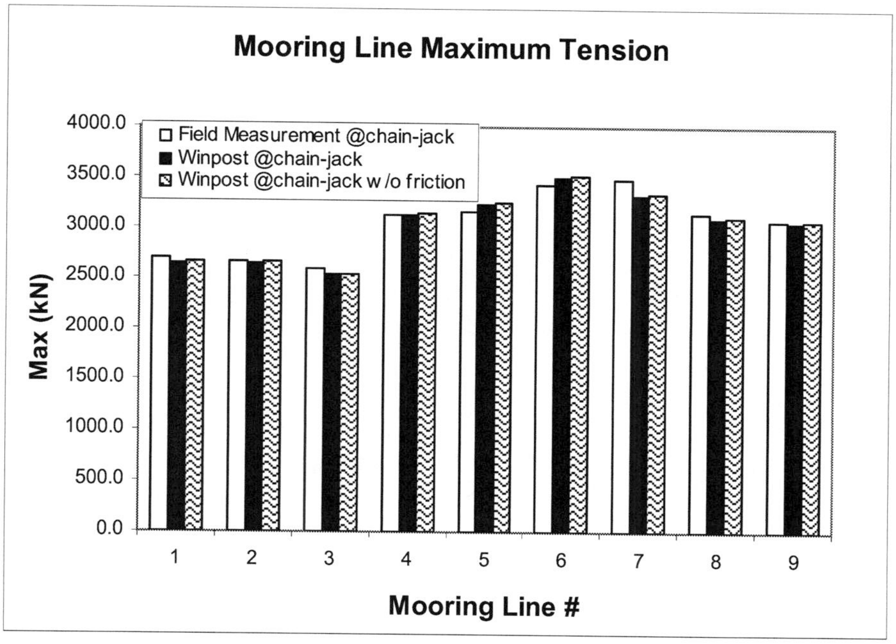

Fig. 9-39. Maximum tension comparison at chain jack with and without frictions, motion from field measurement

Time and Frequency Domain Analysis Versus Field Measurement During Hurricane Isidore

For many years engineers have questioned how accurate the coupled frequency-domain analysis is compared to dynamically coupled time-domain analysis (Kwan and Bruen 1991, Deleuil et al. 1986). Frequency domain analysis has big advantages in computer run times, but the important effect of the phase difference between environmental forces and rigid body motions cannot be captured correctly by this method. Furthermore, linearization of drag forces can result in less accurate results (Ran et al. 1999).

It is worthwhile at this point to compare the results of the two numerical methods— one time domain and the other frequency domain—with field measurements. WINPOST was used to perform analysis for both methods.

To avoid contending with wind direction variability, the wind velocity time series from the field measurement was used as an input to WINPOST. The theoretical JONSWAP spectrum was used for wave input. In the frequency domain, dynamic wind forces were included in the analysis using the spectral method. The wind force time series was transformed into spectrum, taking into account only 20 components of the spectrum for the analysis. Using this method results in 20 percent less wind force standard deviation in surge and sway than using the time domain.

The motion responses presented in this section are for the coordinated system located on the calm free surface. Tension results are from the spar fairlead location (below the chain jack). These analyses do not consider static and dynamic friction.

Figures 9-40 to 9-44 show surge, sway, heave, roll, and pitch comparisons between numerical prediction and field measurement. Figures 9-45 and 9-46 show tension comparisons for mooring lines 1 and 6. Standard deviations for tension have been multiplied by a factor of 10 to clarify the comparison. Figures 9-47 to 9-50 show the low- and wave-frequency density spectrums in surge and roll.

The low-frequency surge responses from the time and frequency domains are slightly lower than the field measurements, while the wave-frequency response is in good agreement. For sway, both the low- and wave-frequency from the time and frequency domains match very well with field measurements.

For low-frequency roll and pitch, the results in the time domain are in good agreement with field measurement, while the results in the frequency domain are overestimated. In the wave-frequency range, results for roll derived using both numerical methods agree with filed measurements, while the results for pitch are slightly overestimated.

Heave results in both the time and frequency domains correlate well with field measurements.

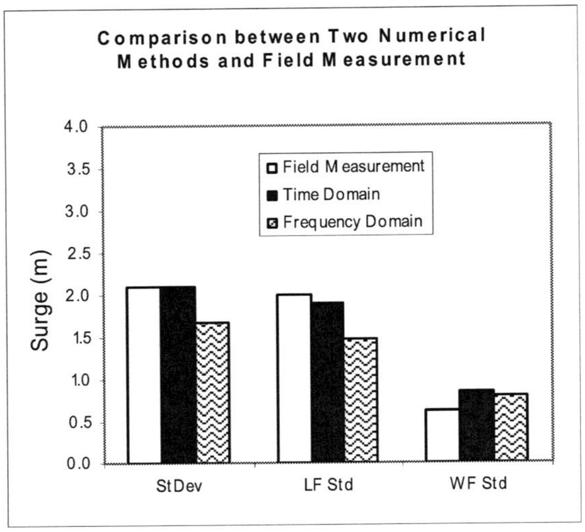

Fig. 9-40. Surge comparison, numerical vs. field measurement

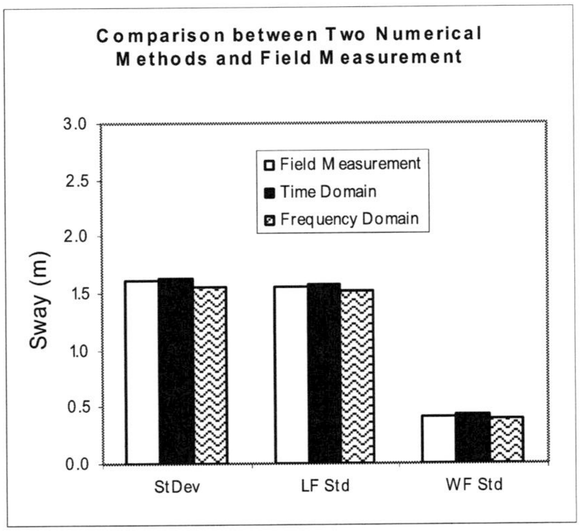

Fig. 9-41. Sway comparison, numerical vs. field measurement

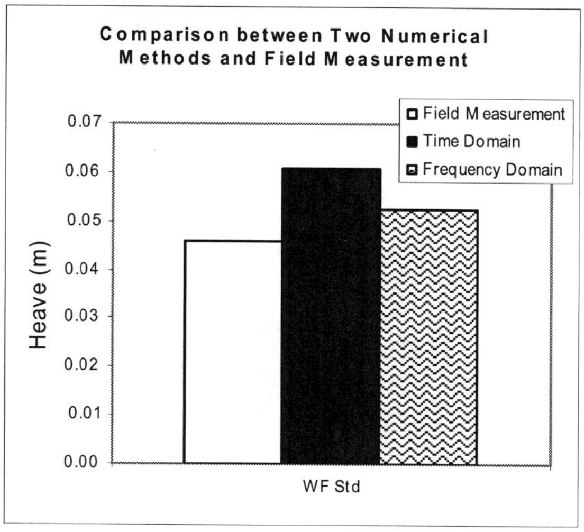

Fig. 9-42. Heave comparison, numerical vs. field measurement

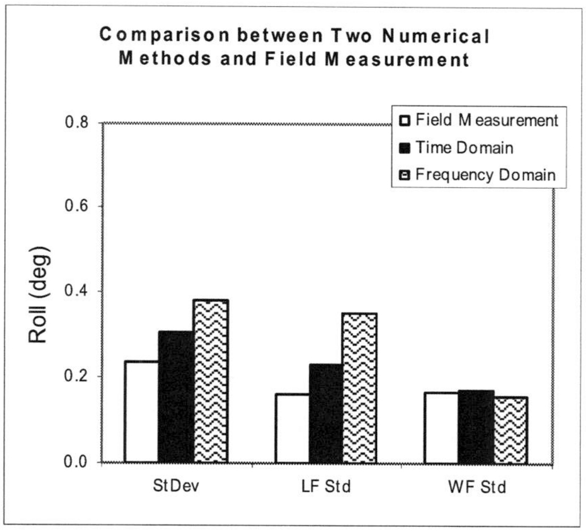

Fig. 9-43. Roll comparison, numerical vs. field measurement

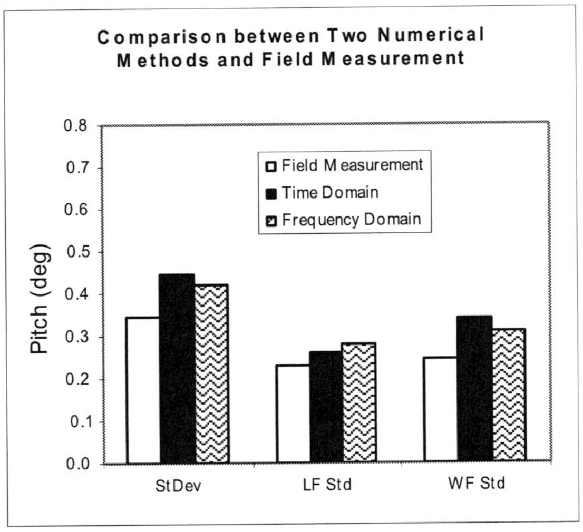

Fig. 9-44. Pitch comparison, numerical vs. field measurement

Although the numerical methods reasonably predict standard deviation for surge and sway, predictions for wave-frequency responses of mooring line tensions turned out to be larger than the field measurement, and the low-frequency responses are in better agreement. The discrepancy in results for the line with the least loaded tension (line 1) is less than that for the most loaded line (line 6).

Discrepancies between numerical analysis and field measurements can be attributed to the uncertainty in the friction coefficients from the fairlead bearing, directionality of the marine environment, and mooring line pretension.

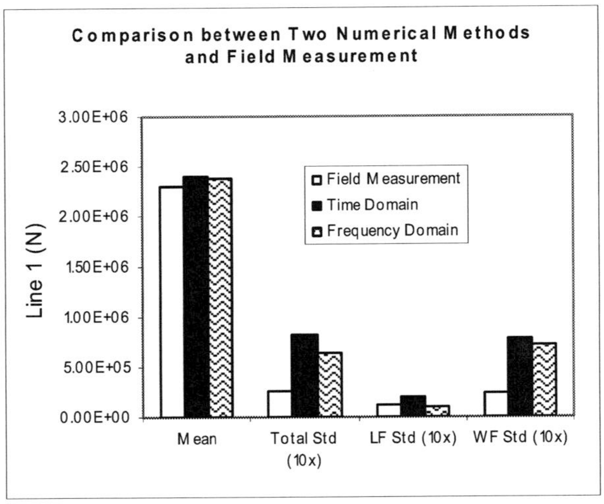

Fig. 9-45. Mooring 1 tension comparison, numerical vs. field measurement

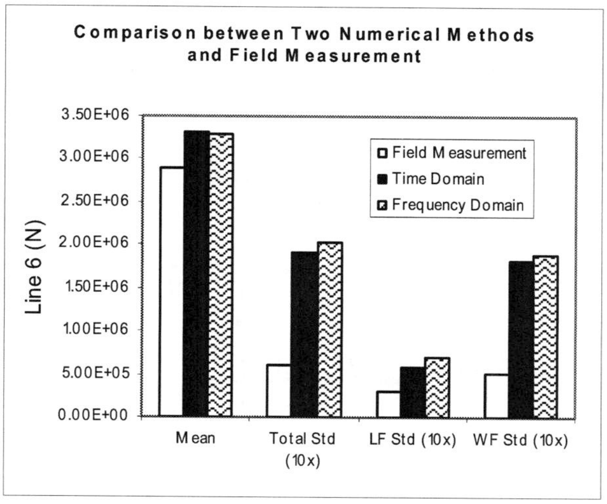

Fig. 9-46. Mooring 6 tension comparison, numerical vs. field measurement

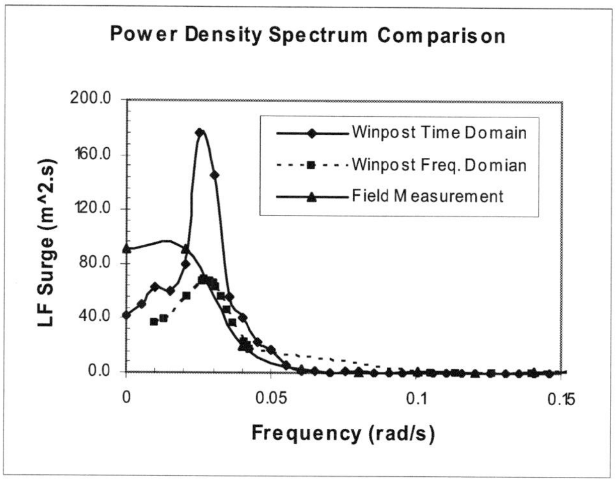

Fig. 9-47. Low-frequency surge spectrum, numerical vs. field measurement

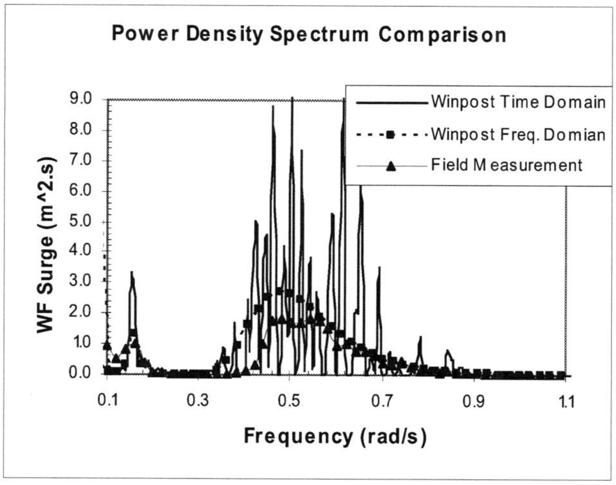

Fig. 9-48. Wave-frequency surge spectrum, numerical vs. field measurement

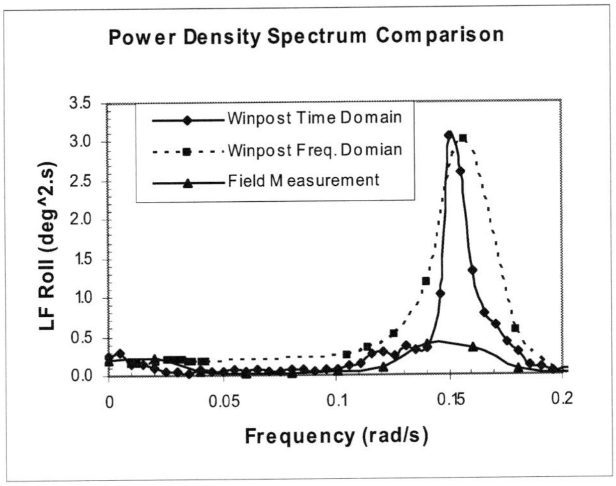

Fig. 9-49. Low-frequency roll spectrum, numerical vs. field measurement

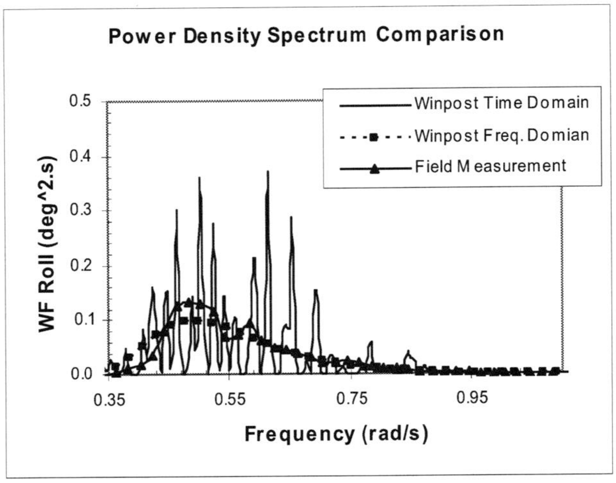

Fig. 9-50. Wave-frequency roll spectrum, numerical vs. field measurement

Summary and Conclusions

Both statically and dynamically coupled time-domain solutions show good agreement with field measurements and yield conservative estimates for pitch and surge accelerations. The data show that dynamically coupled analysis generally provides a closer match with measured motion data. The disadvantage of the approach is that the computer run times for dynamically coupled analysis are greater by a factor of three. For actual design work where numerous cases must be run, the shorter run times and the conservative results of the statically coupled analysis make it an attractive design tool. Dynamically coupled analysis, on the other hand, yields more realistic response estimates, which is significant when evaluating for lower-cost riser solutions.

This analysis computed the Horn Mountain spar's mooring line tensions during Hurricane Isidore using two methods and compared the results with measured tensions. In the first analysis, measured environments were used to predict the spar motions and mooring line tensions, where the environment was assumed to be unidirectional. The second set of simulations used the measured motions to predict line tensions. In both cases, the tensions computed at the fairlead were adjusted to account for fairlead friction and the hanging weight of the chain between the fairlead and the chain jack. This adjustment was necessary because the tensions were measured at the chain jack, not at the outboard location at the fairleads where the computations assume the mooring is attached.

In the first case, standard deviations for tensions are over-predicted by a factor of up to two. Since dynamic tensions are a small component of the total, however, the maximum tensions are predicted to within about 8 percent.

Dynamic tensions for the most loaded lines are still over-predicted, but predictions for most lines are within 10 percent of the field measurements. Maximum tension predictions in these cases are within 1 percent.

Friction in the fairleads appeared to result in about a 10 percent reduction in the dynamic tensions transmitted to the chain jacks. The usual method of computing maximum tensions is to add the weight of the chain above the fairlead to the dynamic tensions of the fairlead computed outboard. The results suggest that this is a conservative method for determining maximum tensions. For a less conservative estimate, fairlead friction could be accounted for in the manner described here.

The analysis also compares the results of using two numerical methods, time and frequency domains, against measured field data for the Horn Mountain spar during Hurricane Isidore. WAMIT was used to compute the first-order wave forces, added mass, and radiation damping in the time and frequency domain analysis. The drag force on the hull and mooring lines was computed from Morison's formula with the assumption that it is proportional to the relative velocity squared. In the time-domain analysis, the drag force was integrated up to the instantaneous free surface at time-varying positions. In the frequency domain analysis, nonlinear drag forces are stochastically linearized, and solutions are obtained by an iterative procedure.

The inability to capture the phase relationship between motion and the exciting force in the frequency domain analysis results in an underestimation of damping near the resonance regions. For that reason, roll and pitch low-frequency responses predicted by the frequency domain are higher than those predicted by the time domain.

The slight underestimate of the low-frequency response in surge and sway could be a result of the lower wind forces in the frequency domain. The responses computed in the frequency domain for the wave frequency range matched well with those calculated in the time domain.

The frequency domain and time domain results for mooring line tensions showed good agreement. Small discrepancies can be attributed in part to the phase difference between hull and mooring line motions, which is not accurately accounted for in the frequency-domain analysis.

When compared to the field measurements, the time-domain analysis shows better agreement than the analysis using the frequency domain. Three factors affect the over-prediction of mooring line tensions—uncertainty in the friction coefficients from the fairlead bearing, directionality of the marine environment, and mooring line pretensions.

Although there is less agreement in results when compared with measured field data, the coupled frequency domain method is adequate for a quick and approximate estimate. The approach certainly offers big advantages in CPU time because it is much faster than the statically coupled time domain method, Frequency-domain analysis can be recommended as tool in pre-front-end engineering design or in a phase where offshore structure design is of an iterative nature.

References

Deleuil, G.E., Des Deserts, C.G. & Shieve, A. (1986). "A New Method for Frequency Domain Analysis of Offshore Structures: Comparison with Time Domain Analysis" *Proc. of OTC*, Paper No. 5303, Houston, TX, USA.

Edwards, R., Shilling, R., Thethi R., & Karayaka, T. (2003). "BP Horn Mountain Spar – Results of Comprehensive Monitoring of Platform and Riser Responses" *Proc. of DOT*, Marseille, France.

Finn, L.D., Maher, J. V., & Gupta, H. (2003). "The Cell Spar and Vortex Induced Vibrations," *Proc. of OTC – Paper No 15244*, Houston, TX, USA.

Garrett, D.L. (1982). "Dynamic Analysis of Slender Rods" *J. of Energy Resources Technology,* 104, 302-307.

Glanville, R. S., Paulling J.R., Halkyard, J.E., & Lehtinen, T. (1991). "Analysis of the Spar Floating Drilling, Production and Storage Structure", *Proc. of OTC – Paper No 6701*, Houston, Texas, USA.

Halkyard, J., Liagre, P. & Tahar, A. (2004). "Full Scale Data Comparison for the Horn Mountain Spar" *Proc. Of OMAE04*, Vancouver, Canada.

Kim, M.H., Tahar, A., & Kim, Y.B. (1999). "Variability of Spar Motion Analysis against Design Methodologies/ Parameters," *Proc. of OMAE'01*, Rio de Janeiro, Brazil.

Kim, M.H & Yue, D.K.P. (1990). "The Complete Second-Order Diffraction Solution for an Axisymmetric Body. Part 2. Bichromatic Incident Waves" *J. of Fluid Mechanics,* 211, 557-593.

Kim, M.H. & Yue, D.K.P. (1991). "Sum- and difference-frequency wave loads on a body in unidirectional Gaussian seas" *J. of Ship Research*, Vol.35, No.2, 127-140.

Kwan, C.T. & Bruen, F.J. (1991). "Mooring Line Dynamics: Comparison of Time Domain, Frequency Domain" *Proc. of OTC*, Paper No. 6657, Houston, TX, USA.

Lee, C.H., Korsmeyer, F.T. (1999). *WAMIT User Manual.* Dept. of Ocean Engineering, MIT.

Newman, J.N. (1974). "Second-order Slowly varying Forces on Vessels in Irregular Waves" *Symp. On Dynamics of Marine Vehicles and Structures in Waves*, London.

Nordgren, R.P. (1974). "On Computation of the Motion of Elastic Rods" *J. of Applied Mechanics,* 41, 777-780.

Prislin, I., Blevins, R.D., & Halkyard, J.E. (1998). "Viscous Damping and Added Mass of Solid Square Plates" *Proc. of OMAE'98*, Lisbon, Portugal.

Ran, Z., & Kim, M.H. (1997). "Nonlinear Coupled Analysis of a Tethered Spar in Waves." *J. of Offshore & Polar Eng.* Vol.7, No.2, 111-118.

Ran, Z., Kim, M.H., & Zheng, W. (1999). "Coupled Dynamic Analysis of a Moored Spar in Random Waves and Currents (Time-Domain Versus Frequency-Domain)", *Transaction of the ASME*, Vol. 121.

Rodenbusch, G., Garrett, D.L. & Anderson, S.L. (1986). "Statistical Linearization of velocity Squared Drag Forces" *Proc. Of OMAE86*, Tokyo, Japan.

Steen, A., Kim, M.H., & Irani, M. (2004). "Prediction of Spar Responses: Model Tests vs. Analysis", *Proc. of OTC – Paper No 16503*, Houston, TX, USA.

Tahar, A., Finn, L., Liagre, P. &. Halkyard, J. (2005). "Full Scale Data Comparison for the Horn Mountain Spar Mooring Line Tensions during Hurricane Isidore" *Proc. Of OMAE05*, Halkidiki, Greece.

Tahar, A., Halkyard, J. & Irani, M. (2006). "Comparison of Time and Frequency Domain with Full Scale Data for the Horn Mountain Spar during Hurricane Isidore." *Proc. Of OMAE06*, Hamburg, Germany.

Index

Page numbers followed by *e*, *f*, and *t* indicate equations, figures, and tables, respectively.